Real-Time Video Content for Virtual Production & Live Entertainment

Real-Time Video Content for Virtual Production & Live Entertainment looks at the evolution of current software and hardware, how these tools are used, and how to plan for productions dependent on real-time content.

From rock concerts to theatre, live television broadcast to film production, art installations to immersive experiences, the book outlines the various applications of real-time video content – the intersection of gaming and performance that is revolutionizing how films are made and how video content is created for screens. Rather than render out a fixed video file, new tools allow for interactive video content that responds to audience activity, camera position, and performer action in real time. Combining software renderers with environmental information, video content is generated nearly instantaneously to simulate depth, creating a new world of Virtual Production. This book provides an overview of the current software and hardware used to create real-time content while also reviewing the various external technologies the real-time content is dependent upon. Case studies from industry experts appear in each chapter to reinforce the tools described, establish industry practice, and provide insight on a complex and rapidly growing discipline.

Real-Time Video Content for Virtual Production & Live Entertainment prepares students and practitioners for a future working with real-time technologies and informs current entertainment technology professionals how to rethink their old roles using these new tools.

The book includes access to a companion website featuring web-based and video resources that expand on topics covered in the text. Each chapter has a unique page that points to example material, video presentations, and professional studies on chapter topics. You can visit the companion website at rtv-book.com.

Laura Frank is a creator advocate and founder of frame:work, a professional organization for the creative video community. She spent her early career as a lighting programmer with projects spanning rock tours with David Bowie and Madonna, to Broadway shows like *Spamalot* and television events like the *Concert for New York City*. Laura made the shift to screen content and control with the growth of the media server market. After a decade of refining a media delivery workflow, she established herself as a Screens Producer leading a highly regarded Media Operations team for prominent events around the world. Her shows included the MTV Video Music Awards, The Game Awards, the Turner Upfront, and the CMA Music Awards.

Real-Time Video Content for Virtual Production & Live Entertainment

A Learning Roadmap for an Evolving Practice

Laura Frank

NEW YORK AND LONDON

Designed cover image: Inside the Interactive Experience at the Al Wasl Plaza, Dubai, 2022. Image Credit: Bild Studios

First published 2023
by Routledge
605 Third Avenue, New York, NY 10158

and by Routledge
4 Park Square, Milton Park, Abingdon, Oxon, OX14 4RN

Routledge is an imprint of the Taylor & Francis Group, an informa business

Library of Congress Cataloging-in-Publication Data
Names: Frank, Laura, author.
Title: Real-time video content for virtual production & live entertainment: A learning roadmap for an evolving practice / Laura Frank.
Description: New York : Routledge, 2023. | Includes index.
Identifiers: LCCN 2022034077 (print) | LCCN 2022034078 (ebook) | ISBN 9781032073477 (hardback) | ISBN 9781032073446 (paperback) | ISBN 9781003206491 (ebook)
Subjects: LCSH: Performing arts—Production and direction. | Digital video. | Real-time data processing.
Classification: LCC PN1590.P74 F73 2023 (print) | LCC PN1590.P74 (ebook) | DDC 791—dc23/eng/20220930
LC record available at https://lccn.loc.gov/2022034077
LC ebook record available at https://lccn.loc.gov/2022034078

ISBN: 978-1-032-07347-7 (hbk)
ISBN: 978-1-032-07344-6 (pbk)
ISBN: 978-1-003-20649-1 (ebk)

DOI: 10.4324/9781003206491

Typeset in ITC Officina Sans
by Apex CoVantage, LLC

Access the companion website: www.rtv-book.com

Contents

APPENDICES

Contributors

Case Studies

Jake Alexander, LuxMC
Project: The League of Legends World Championship, Shanghai, CN, 2020

David Bajt, Bild Studios
Project: Al Wasl – The Interactive Experience, Dubai, AE, 2022

Miyu Burns, Rich & Miyu
Project: AR Livestreamer, Tokyo, JP, 2021

Richard Burns, Rich & Miyu
Project: AR Livestreamer, Tokyo, JP, 2021

Charli Davis, Luke Halls Studio
Project: Forest of Us, Miami, US, 2021

Danny Firpo, All Of It Now
Project: Coldplay X BTS, "My Universe" on *The Voice*, Los Angeles, US, 2022

Fernando González Ortiz
Project: *The Lion and the Firebird*, New York City, US, 2022

Alexandra Hartman, Savages
Project: Studio Growth Project, Los Angeles, 2022

Kerstin Hovland, Electronic Countermeasures
Project: The Foo Fighters 26th Anniversary Tour, US, 2021

Fiona Jennison, for Bluman Associates
Project: Dave at The BRITs, London, GB, 2020

Kristaps Liseks, Evoke Studios
Project: Siemens Hannover Messe Digital 2021, Bavaria, DE

Emery Martin, Electronic Countermeasures
Project: The Foo Fighters 26th Anniversary Tour, US, 2021

Chema Menendez, Evoke Studios
Project: Siemens Hannover Messe Digital 2021, Bavaria, DE

Blair Neal
Project: Trace Your Road, Rome, IT, 2013

Urs Nyffenegger, Evoke Studios
Project: Siemens Hannover Messe Digital 2021, Bavaria, DE

Nils Porrmann, dandelion & burdock
Project: Johnnie Walker Princes Street Interactive World Map, Edinburgh, GB, 2021

Finn Ross, FRAY Studio
Project: *Back to the Future the Musical*, London, GB, 2021

Vincent Steenhoek, Evoke Studios
Project: Siemens Hannover Messe Digital 2021, Bavaria, DE

Emily Zamber, Visual Noise Creative
Project: MTV Video Music Awards, Los Angeles, US, 2021

Quotes

Emric Epstein, VYV
Martin Granger-Piche, VYV
Greg Hermanovic, Derivative
Francis Maes, SMODE TECH
Ashraf Nehru, disguise
Matt Swoboda, Notch

Acknowledgments

This textbook contains a snapshot of a moment in time. History is not a story easily told as it is happening, yet I believe the early 2020s will be remembered as the starting point of an evolutionary leap of entertainment technologies that were accelerated by the pandemic. Combined together and dependent on real-time content creation, this is where we go off the map, and chart a new path forward. This history is still unfolding.

The technologies discussed are not always new; the combination and application of technologies represented here are what is new. The results inspire further advancement in these tools at an incredible pace. I am grateful for the many, many discussions I've had with my contributors and other industry practitioners to help me see that advance unfold as it occurred. It has been a bumpy ride for all of us. These insights were hard earned.

First, I have to thank Ashraf Nehru. While his work reshaped my professional career, it was a conversation about the future of our industry that inspired this book. I tease him when I'm frustrated with writing a history of the future that my struggles are his fault. In truth, I am grateful.

My sincere thanks to all the contributors to this textbook. There is an incredible amount of work and expertise shared in the case studies that were written with great care. I know time is at a premium and the quality of work that was shared is humbling. I encourage readers to visit the online version of the List of Contributors where you can find links to their companies, other works, as well as other case studies.

There are many others I am indebted to for their time and expertise. Some wrote case studies that I was not able to incorporate into the final textbook. Josue Ibañez and Bris Pineda from Cocolab, Tito Sabatini from Duo2, Andrey Yamkovoy and Yuri Kostenko from Front Pictures, Gil Castro at Intus, Juliette Buffard at Superbien, and Greg Russell at Xite Labs are included on the companion website and I encourage you to take a look at their work.

Sarah Cox (Neutral Human), Zach Alexander (LuxMC), and Phil Galler (LuxMC), were gracious enough to spend time debating terminology with me. JT Rooney (XR Studios), Simon Anaya (AnayaFX), and Nick Rivero (Meptik) kindly supplied images that I needed. Ben Nicholson (Miami University), Matthew Ward (Superlumenary), and Trevor Burk (VNC) assisted me with early outlines. My thanks to all of you.

I must also express my immense gratitude to Ian MacIntosh who created images for the textbook. Ian has been supporting my community project, frame:work, for the last year and produced many of the images seen in chapters 7 and 11. He is a thoughtful and talented artist and creative technologist and I wish him all the success in his future endeavors. My thanks to Sharon Huizinga (CCM) for introducing us and also providing me with regular insights into life and work.

My thanks to the team at Routledge for their assistance, Stacey Walker and Lucia Accorsi. Thanks again to Matthew Ward who served as technical editor and provided valuable insight shared in this book as well as on the companion website.

Finally, I am incredibly grateful for my husband, Daniel Damkoehler, who read every word of this book and helped me craft its meaning and delivery. He watched me seek to capture, organize, and write this book, and supported me through each chapter. I could not have completed this work without his help. Thanks, lovey.

Laura
Summer 2022

CHAPTER 1
Introduction

We Are Surrounded by Visual Content

Video screens dominate our daily life. We carry small video screens in our pockets, used so often we monitor our screen time. Mid-size screens are mounted centrally in every home, restaurant, and waiting room. We pay to watch large screens in a movie theater with a group of strangers. When we step outside, building-size screens blast us with information at traffic intersections and city centers.

The content on these screens is used to entertain us with stories or games, engage us visually with art, or deliver information. Much of our screen time is devoted to information gathering on the internet, a display environment fully responsive to user input. The imagery displayed on a web page instantly responds to the person interacting with the screen. When you go to the page of an item for sale at a large online retailer, you see one product out of an inventory of tens of thousands of items. When you visit your favorite online video service you are served a list of suggested content specific to your user profile. It would be impossible to build and maintain a unique web page for each item sold or each YouTube file. Instead, data and code are dynamically rendered to become the image content of browser windows and app views.

Video responds to us outside the world of the internet. Perhaps you've experienced an art installation with interactive content. Devices sense your position and movement as you move through the installation. One part of the exhibit might generate a path of sparkles along the walls in response to your arm waving and another area renders cracks in the floor under each footstep. Large immersive video installations of artists are popping up in cities all over the world. Viewers of these exhibits are more sophisticated in their expectations of how the imagery will look and respond to their presence.

While these are examples from the internet and art world, we describe these on-demand visual displays using a set of terms that are now common to the creative video professional. Content is *generated*. Code is *rendered*. What once fell within the domain of interactive content now belongs to the world of creative video content production. Interactive capabilities common to the web and video art are ideal for live performance and the new field of Virtual Production. Creative video production is shifting to tools made for real-time content creation.

Creative Video for Live Visuals

The use of creative video in live entertainment is considered to have a relatively short history. Projection Design for the theater is dependent on the development of projector technology after all. Pioneers like Erwin Piscator and Josef Svoboda used film projectors in their stage designs in the 1920s and 1950s respectively. However, real-time content creation use in entertainment has a rich analog history. Asian shadow puppet theater goes back thousands of years and requires only a candle, cut-out characters and puppeteers to populate a screen full of imagery.

One of the earliest contemporary forms of projected real-time content creation dates back to the late 1960s. The Joshua Light Show used many components, most famously "liquid light," a combination of colored oils projected onto a surface, to create visual moods for bands performing at the pioneering rock venue, The Fillmore East. The designer mixed the oils to create a randomized visual experience one might call the granddaddy of VJ'ing. While the Joshua Light Show carries on to this day, what is interesting is the five decades long road we took from analog generated visuals to today's digital solutions for real-time content creation.

Check out rtv-book.com/chapter1 for video examples of the Joshua Light Show and Shadow Puppets.

DOI: 10.4324/9781003206491-1

The projectors and analog overhead lamps and lenses used in the Joshua Light Show (and many high school AV departments) were gradually replaced by laser driven digital projectors outputting 40K lumens. Slide projectors, PANI and PIJI, RGB lamp, and the Texas Instruments DLP mirror chip were all engineering steps along the way.

All night work calls to align projectors is being replaced by structured-light-driven automated projector calibration. LED screens have developed from 25mm pixel pitch tri-color modules to 1.2mm pixel pitch sources with red, green, and blue packaged together. At the same time, computer graphics processing has advanced from rendering sub-kilobyte grayscale images to real-time rendered high resolution motion graphics.

Most relevant to the practice of stage design, painted scenic walls combined with a few video screens covering 10% or less of a performance stage have been replaced by sets made only with video screens. This might seem reasonable to accomplish in 60 years, but most of this growth occurred in the last 15 years. Like so many periods of rapid development, practitioners are running to catch up with the technology. We have yet to establish a consistent set of workflows, team structures, vocabulary, best practices, and communication tools that are common to other entertainment design practices.

There are many unique requirements for video content in performance that would benefit from shared practice. As with our examples of how we use screens in daily life, the content in performance can be narrative, creative, interactive, or informational, each with distinct production workflows. The visuals on the screen can tell the story of a pop song or subtly contextualize the background of a CEO's presentation. At other times, the visuals are more informational, representing infographics designed both to entertain and provide insight with text. In live performance, visuals extend the scenic design, replacing static walls with dynamic surfaces that animate and transform the visual environment for audience and performer.

In my work as a screens producer, I often recommend that scenic screen content creators take time to study scenic design. No matter the content type, the principles of set design are invaluable to creative video production and too often overlooked. Even simple moving visuals can upstage compelling performers and presenters. The audience's eyes naturally go to the most active, the brightest, or the largest image on stage. As screen real estate overtakes more of the set, it can eclipse the reason the audience came in the first place: the performers.

This text will not cover the principles of projection design or scenic design. **Our focus is on the use of real-time content creation and the collaboration of screens with all of the other performance design disciplines:** lighting, set, sound, camera, and the performers themselves. The addition of tools that allow for real-time manipulation of the imagery on those screens can enrich and invigorate this collaboration in profound and new ways.

Performance is a collaborative art form between director, actors, and the design and technical teams. Video screens are a contributor to that collaboration, though historically, in a fixed state once a production is locked. Projection used in live performance began with physical film and slides, media that is unchangeable once "printed." When video playback moved to hard drives, media servers offered a collaboration opportunity for minor adjustments to color, playback speed and direction, as well as rudimentary real-time effects.

Media servers came about much like the Joshua Light Show, in response to a desire for exciting, flexible music driven visuals. Even in basic form, effects engines in media servers represented a paradigm shift in creative flexibility for screen design. Other digital playback devices common in the AV and broadcast world can do no more than organize and output video files for playback. The appearance of media servers coincided with a key development in screen display technology: a growing market of affordable creative LED technology.

Media server effects engines made video an active participant in performance collaboration. Video screens were no longer static, pre-planned, expensive to change, visual elements. Screens became a dynamic complement to the flexibility of the moving light rig, often managed by the same creative programmers. Media servers also became an important part of the DJ market, adding a visual layer to the playback performance experience and creating the live VJ role.

Video effects engines also opened a door to real-time content manipulation. The teams that manage these tools are creative partners with the performers and traditional entertainment design teams. As screens overtook the physical design space of productions, often replacing space that lighting rigs or fixed scenery once occupied, demand increased for customized creative video screen content. However, custom content has a liability. Creative changes beyond minor adjustments require technical edits and rendering time.

Creative video content creators are well acquainted with the many challenges of rendering edits under time pressure, usually resulting in long hours and lost sleep. These hours become harder to explain to a generation of producers raised on phones that can track their faces with cartoon features in real time. Is there a way we can make this process better? Is real-time content creation the path forward? It can be.

In this textbook, we will review the evolution of video from rendered "static" pixels to code generated real-time visuals, a set of changes that revolutionized the collaborative capabilities of creative video design. These advances allow video designers to respond in the moment to client requests, performing artist demands, and unpredictable environmental inputs. Real-time content creation has forever changed the landscape of how we use screens in performance.

However, the real-time content evolution came shockingly fast to a screens design discipline barely out of its infancy. The real-time video content working environment provides near infinite flexibility, which can be a blessing and a curse. Great visual imagery comes from thoughtfully designed and executed creative direction. Because of the perceived ease of real-time manipulation and flexibility, an unfortunate negative practice has emerged as a side effect of all these gains, the tendency for production teams to push essential decisions to an undefined "later date."

Real-time video content design practices allow for flexibility but should always start with design planning and creative discussion, as with any performance design practice. Possibly the best analogy would be a jazz music performance. As collaborators, these musicians must rehearse extensively to succeed together as collaborative improvisers. Rehearsal prepares the musicians to not only excel at their instrument, but practice listening to each other. Combined together, this allows the musicians to improvise in real time. In the same way that mastering musical structures, rhythms, and aural themes prepares jazz musicians to improvise in performance, real-time video content professionals must prepare visual structures, themes, and variables so they can effectively reinvent their work once onsite.

Live Entertainment Is Forever Changed

We can't talk about real-time video content without talking about the disruption of the Covid-19 pandemic to the live entertainment industry. In March of 2020, almost all live audience entertainment came to a halt. All over the world, social distancing measures required the closure of movie theaters, concert halls, and theaters. Countless thousands of entertainment professionals were out of work overnight.

Only a few years had passed since real-time generated digital content was introduced to live entertainment screens production. In the mid-2010s, Notch, a real-time content creation tool, was in beta and many content design teams started experimenting with real-time features to content creation. Before that, custom code based generated content was primarily used for art projects, installations, and other environments that required responsive imagery and weren't limited by short production schedules. Around the same time, world-building game development tools like Unity and Unreal were repurposed as entertainment production pre-visualization tools, primarily for film.

When the pandemic hit live entertainment, the media server and creative video community continued developing their real-time toolsets. These solutions presented an opportunity to get back to work using real-time generated content and deliver live shows on the internet, in virtual environments. "Virtual events" became the entertainment buzzword of summer 2020.

Prior to the pandemic, production teams experimented with real-time content tools and Augmented Reality, or AR. To be effective, the media server system needed to "know" where the camera was in space and where it was looking, as well as what it looked at. We will examine this technology in more detail in later chapters.

> Check out rtv-book.com/chapter1 for an early example of AR used in live television broadcast in 2010 for the Video Game Awards

While many in the real-time content field have explored various approaches to Augmented Reality over the last two decades, no single solution rose to become common use except for sports broadcast applications. In that market, AR found a home displaying game statistics, expanding in complexity over the last 30 years to include on-field graphics and host of other features that sports fans have come to expect. Concurrently, video screens were finding use on film sets. LED screens have become a popular lighting source over the last ten years for creating natural looking reflections on scenic surfaces. Pre-pandemic, a few pioneering filmmakers used high resolution LED or projection as a dynamic background while shooting on a soundstage.

> Check out rtv-book.com/chapter1 for a video explainer of the projection backgrounds used on the film Oblivion from 2013

There's no way to say how long real-time content tools like Notch, Unreal, and Unity would have taken to dominate the production industry if it were not for the disruption of the pandemic. Notch was increasingly popular in live entertainment content production. Unreal and Unity both had a strong foothold in the film VFX pipeline. Media servers were used for film background and reflection screens. All of these solutions needed refinement in order to become effective when used together.

With so many screen technologists and creators at home in March 2020, prior experiments combining AR with camera perspective rendered dynamic backgrounds became an industry focus. Live production and filmed entertainment needed to pivot to get back to work, and these tools looked like the best path forward. With much of the world stuck at home, demand for online entertainment content soared, accelerating Virtual Production technology development and adoption. As the pandemic restrictions subside, it's anyone's guess what the future of Covid-19 will be, but it is clear Virtual Production tools are here to stay.

> Virtual Production is an umbrella term to describe an ever-growing range of camera dependent digital production tools including Augmented Reality (AR), In-Camera Visual Effects (ICVFX), Background Replacement technology, Extended Reality (XR), Mixed Reality (MR), scenic extension, and more. We will explore and define these terms in detail in future chapters.

The pandemic has forever changed entertainment production. Aside from accelerating the emergence of Virtual Production, the pandemic forced the adoption of a plethora of remote work practices, most of them new to the entertainment industry. Audiences went fully remote and the consensus is that in the future we will see more in-person + remote hybrid audiences.

Like everyone else, performers adopted ways of working remotely in the pandemic, using the web to deliver everything from a phone camera feed in their living room to a robotic broadcast camera feed from a studio. Production teams worked remote too, running lighting rigs by piping Artnet over the internet, connecting via video calls to coordinate camera feeds of distant stages, and build media servers in the cloud.

Our remote capabilities put screen technology front and center as everyone spent hours a day on their screens scheming new ways to meet and entertain one another online. The world looked to the wider screens community to drive the technology to make entertainment online a better experience. All of us have seen some exciting examples demonstrating the potential of virtual events, but we are still only a few steps along a long path to discovering the potential of what we can create with these tools. This new world of screen savvy connected audiences and power visual tools offers a wide range of interesting new career pathways.

Real-Time Video Content Is the Engine that Powers Virtual Production

We need not look hard to find many film and broadcast approaches to Virtual Production. Some studios operate with nothing but LED walls, called a "Volume," and rely on a direct relationship between the camera and a game engine to drive the visuals in real time. Smaller LED stages include an LED floor and rely on real-time software to extend the visuals beyond the edges of the LED walls. Other solutions eliminate the video walls, opting for the ease and cost effectiveness of green screen. Yes, that green screen, the one that has been around since the 1940s and dominated cable access studios throughout the 1980s.

> Check out rtv-book.com/chapter1 for a video explainer outlining the progression of LED screen from green screen on *The Mandalorian* in 2019

None of what makes Virtual Production possible could exist without the dynamic relationship of the camera information driving real-time content creation to the devices that process its signal to screen. Real-time content depends on processing power, but also on the highly skilled teams that leverage that processing power to deliver the desired results. There's a reason we originally referred to the film team managing a real-time content delivered to the screens of a Volume as the "Brain Bar." Thankfully that term is being replaced with clearer terms like "Technical Art Department" or "VAD (Virtual Art Department) Operations."

VP will continue to change based on the continued improvement of graphics cards and the software that generates content in real time. Think of the first home video games and how they have improved in the intervening 40+ years. My first exposure to real-time graphics was the Pong game my grandfather brought home in the mid-1970s. All that game console needed to do was draw white lines and a square ball controlled by the user defined movement of the paddles. Today, games render tens of thousands of polygons multiple times a second to create immersive, near movie CGI quality visuals.

What we are building now in the entertainment community is the foundation of an incredible field of real-time content production.

And while Virtual Production is one exciting result, many more developments are yet to come. Our production community is on the cusp of delivering the creative, engineering, and operational capacity to drive the future of the Metaverse.

What This Book Is Not

For every application of real-time tools, there are groups of people innovating unique protocols and practices to use these tools. The discipline is changing as I write these words and will make several major leaps by the time you read these pages.

So let's talk about what this textbook is not. This work will not teach you the specific tools of real-time content creation. This book is not a "how to" of Virtual Production.

Whether your interest is in Extended Reality, previsualization, live concert creative video, Augmented Reality, creative video scenery, virtual events, interactive installations, Mixed Reality, LED stages, 3D modeling, In Camera VFX, or any of the many other digitally manipulated visual experiences that make up generated video immersion entertainment, the best learning is doing. Your best teachers are each other and the challenges you take on.

This book *will* tell you how we got here and how to think through your approach to real-time content creation. This book *will* help you become a successful production partner, engaged team member and career-long learner. Students of entertainment technology interested in real-time content production as well as professionals who want to move into this field will benefit from the discussion within these pages.

SECTION 1

Exploring Real-Time Content

CHAPTER 2

Defining Real-Time Video Content

Defining the Term "Real-Time"

What do we mean by the descriptor "real-time?" Isn't everything we do happening in some real amount of time? When is time not real? If you've ever waited on a video file to render or experienced a video server struggle to play at "normal" speed, this is what we mean in the video world by real-time: the playback of video information accurately at the intended frame rate.

References to time are common to video communication. Video can be slow-motion where time is stretched relative to actual speed. Or video can be hi-speed, where time is compressed. These terms are typical of shot footage as they represent a comparison of playback speed to the speed an event occurred when captured. But what if the event isn't captured? Instead, what if the event is simulated or created with a computer? What defines time then?

"Real-time" became a common buzzword in many industries to describe the time necessary to complete a simulation, a computed representation, of something happening in the real world. Stock prices and money transactions, web page interactions and social media "likes," all need to be captured and reported upon as they occur. When we discuss a computer simulation of real world data, we have to talk about the time needed to compute the result of that model relative to real world time. Does the data interpreted account for a minute of the real world, or a millisecond? Can it be processed in the same or less time than it represents in playback? That is a standard computer science definition of real-time: the computation introduces no delay versus events in the real world.

Our experience of the world happens in real time. We have divided the flow of that time into seconds, minutes, hours, days, weeks, months, years. The measuring sticks of time are based on principles of physics. One rock trapped in a very long orbit around another larger rock makes a year. One subatomic particle's behavior within a Cesium-133 atom defines a second. We don't think of time as having a speed, but as mammals we are highly sensitive to when the time of the world does not align with our expectations. Try moving in slow motion in front of a dog; it freaks them out.

Where we most often encounter that sense of real-time at the wrong speed is when we try to describe the real world with a computer. Within a computer, speed depends on processing power and time is a user defined formula of how often to deliver results. The complexity of the result we want relative to the capabilities of the computer we are using will determine how long we are watching progress bars and spinning wheels. If we want 30 images every second, our computer needs to process and output 30 images, formulas, operations, and data within the one second window.

If you have ever sat watching the progress bar of your content creation software render out a video, you already appreciate the meaning and significance of the term "real-time." Have you waited 30 minutes for a one minute clip to render? Wouldn't it be much more productive to wait one minute for a one minute video clip to render? In that case, you could play the frames as they render and at the exact moment that you need them.

We can describe real-time video as video that is rendered at the same time its video frames are visible to the viewer. A one-minute video clip will take one minute to render. The system processes the individual frames in the same instant they are projected or displayed.

Like a computer model or simulation, a real-time video file is composed of a set of instructions. When a system processes those instructions, the result is a sequence of images at a specified frame rate. A rendered video file is a containerized sequence of images. Once rendered, a video file plays back in real time, to a specified frame rate. So what's the difference between the two?

DOI: 10.4324/9781003206491-3

The difference is functionality. A real-time video file is always available to be played back since the system will render it as we view it. The rendered video file requires two steps, rendering and viewing. The rendering period can take much longer than the actual viewing time. Real-time video content is made possible by the availability of the file structure and computing power to play files in real time and produce quality results. By combining the render step with playback, we can incorporate interactive image manipulation and creatively augment files and instantly review the results.

Let's say we have a simple set of instructions. We want to draw a rotating cube in an empty space. Our instructions to the processor must describe both the shape and its movement. Any shape, in this example a cube, can be mathematically described (see Figure 2.1). Movement, in this case, rotation of the cube, can be described by a vector instruction for motion. Most modern computer processing could easily draw hundreds of frames per second of this example. However, there was a time when real-time rendering of a rotating cube was a significant computational challenge.

Visit rtv-book.com/chapter2 for high resolution reference figures as well as supporting video.

As computer capabilities advanced, real-time rendering of a complex model with multiple thousands of polygons, lighting, reflections, and material properties was possible. Many computers today would still struggle to draw the maximum level of detail shown in Figure 2.2 at 30 frames per second. To remain real, or accurately tied to the correct time, there are two options: draw as many frames as you can to describe each second of the content, or reduce the complexity of the instruction set to meet the processing capacity of the hardware used at the requested frame rate. Video real-time capabilities are always governed by processing power and the desired visual result.

People often refer to real-time content as unrendered. This is not technically correct. As described earlier in the chapter, the act of playing real-time video means you are rendering the content, creating each frame, on demand. If that demand is for a frame every 30th of a second, your instruction set for the real-time file will render each frame at the specified frame rate. This on-demand rendering is what we consider real-time content. The file exists before playback in an unrendered state, but is rendered upon display.

Often what people mean by "unrendered" is the fact that a real-time video file exists on the playback system as code. The host computer reads and executes that code to nearly simultaneously render and display the required frames. One key advantage of delivering an executable file is significant file size reduction versus a rendered alternative. Reduced size means transferring or updating files is easier and faster. For example, an unrendered file representing 20 minutes of video content could be 1/10 or smaller than 20 minutes of rendered video.

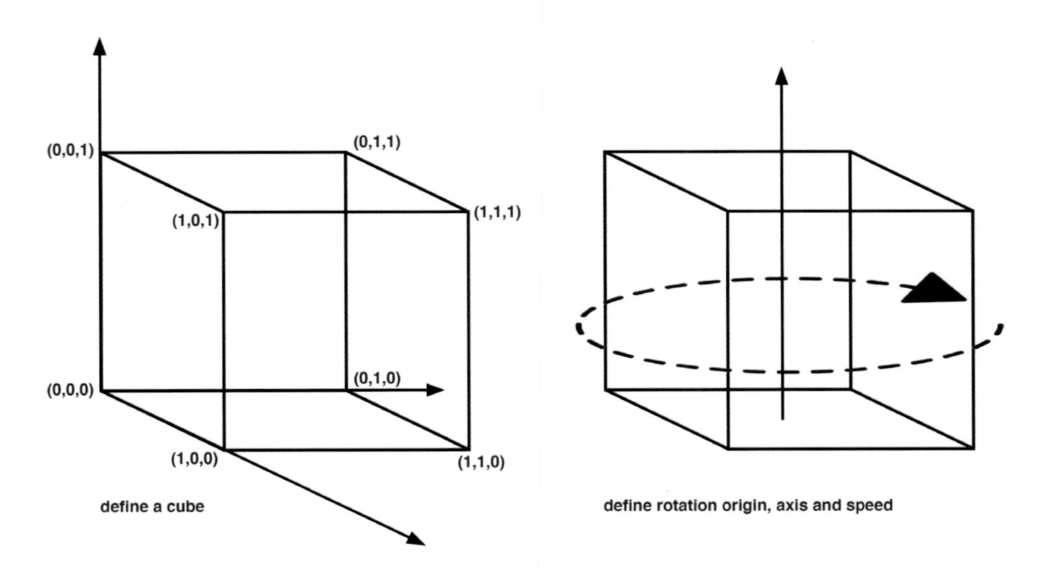

Figure 2.1 A simple rendering task for a modern computer. Source: Image by Author.

Figure 2.2 Increasing complexity of 3D cityscape models. Source: Image Credit: iStock.com/jamesteohart and iStock.com/cherezoff.

Imagine you need to make a small edit in the middle of a 20 minute file. This edit will affect only 10 seconds of the entire file about 5 minutes from the beginning. For content produced by a traditional content creation software application, after completing the edit, we would need to re-render the entire 20 minute file. When using a real-time software application, once edited, a new executable file is immediately available for viewing.

Sometimes, we can edit content within the software platform hosting the real-time file, making this process even more efficient. A content designer can incorporate specific creative variables in anticipation of the need for quick updates. These variables allow for significant creative flexibility and quick image manipulation.

Users of media servers as video playback tools commonly alter the color of video files during playback for creative augmentation of an existing file. The video file may be rendered, but the media server can adjust color in real time during playback without negatively affecting speed or other aspects of the content. In this example, color is an "effects" variable the media server can control. Other effects manipulation of rendered video files requires heavier processing and will slow down the frame rate or drop frames to maintain playback speed.

For real-time files, the content design can include variables that the media server can alter as the file is "playing" or executed. Overall color can be a variable, as with a hue shift, but designers may add many more levels of detail. We can apply a new color to a specific part of the real-time file content, instead of shading the entire rendered frame. In Figure 2.3, the middle image shows a media server style color alteration that changes the full image, while the image on the right shows the result of a real-time file with a color variable applied only to the clouds.

Real-time video content variables must be part of the design process and planned for to be effective. However, we can't anticipate every possible variable. Just as there are rules and limits within a video game, there are limits to what we can accomplish with a real-time video content file based on its original design. Anything is possible, but new ideas need time, discussion, and usually will impact the budget.

Creative flexibility is one of the driving forces for the adoption of real-time content creation. Traditionally rendered content is fixed, printed like a document. In Figure 2.3, we've reviewed the limitations of media server manipulation of a rendered file. Media server effects are typically whole frame alterations. With a real-time file, that image could have variables for not only cloud color, but also the number of clouds, their size, their distance from the ground, and more. Introducing variables into real-time executable files enables detailed manipulation of the image results.

Variable overload: don't let your creative process get bogged down in anticipating potentially needed variables. Some creative needs arise as discoveries while collaborating on the project. There is no way to plan for every possibility.

Figure 2.3 Example of full frame color effects vs real-time variable only applied to clouds. Source: Image Credit: iStock.com/ONYXprj and Author.

Typically, rendered content will require content creators to go back to the content creation software and make edits to accommodate design changes. This can be time-consuming, both in terms of effort required to update the video content and video rendering (or re-rendering) time. Well planned real-time content preparation can be time consuming. But with careful and thoughtful creative development, a real-time video file will contain enough flexibility to enable quick changes. This eliminates the need for re-rendering content and saves valuable time.

However, complex changes and re-designs will take time regardless of the style of content creation. Real-time can unfortunately be misunderstood to mean "last minute." This is never the case. When using real-time content for an event, I always explain to clients that we need to allow the same amount of preparation time as we would for any traditional content creation solution. Planning, storyboarding, mood boards, creative consultation: these are all critical to creating good content design. Too often teams show up onsite expecting to have infinite possibility because they assume real-time equals infinite capabilities.

We all know the proverb (popularized in *Spider Man*), "with great power comes great responsibility." Real-time content creation tools are powerful, but that means they come with the responsibility for a near-infinite number of choices. As with any design, great creative comes with communication, planning, and development time. Even though the content no longer requires lengthy rendering times, that doesn't mean that good content doesn't require lead time to plan and design. Real-time content time tools cannot magically create content from nothing.

The Master Variable: Time

As we've described, the number of frames a real-time content file generates per second depends on two factors: hardware limitations (usually the capability of the computer or media server's graphics card), and image complexity per frame (this is the rendering or technical as opposed to visual complexity of the imagery). Time is fluid until the user defines time in terms of frames per second. For this textbook, we will use the US Standard of 30 frames per second (30fps).

We've now established a 30fps clock, described visually as a timeline. Let's imagine an example of a 3 minute long video file contained in an 85kb executable file. This file exists along the timeline, limited to filling in a space 3 minutes long (see Figure 2.4). While the script might be simple enough that the server can produce more than 30fps, the script playback will remain contained within this defined time span. The server will discard extra frames if we run long or repeat frames to fill the cycles of time required to compensate for each second of any shortfall.

However, time is also a variable. We can slow down or speed up the playback of any video clip as well as the speed of any real-time file (see Figure 2.5).

In playback control, we can trigger specific events during file playback. A command can change the size and shape of elements in a video clip or introduce new visual elements. We refer to these events as cues (see Figure 2.6). With a cue, we can change the value of a parameter while the file is playing.

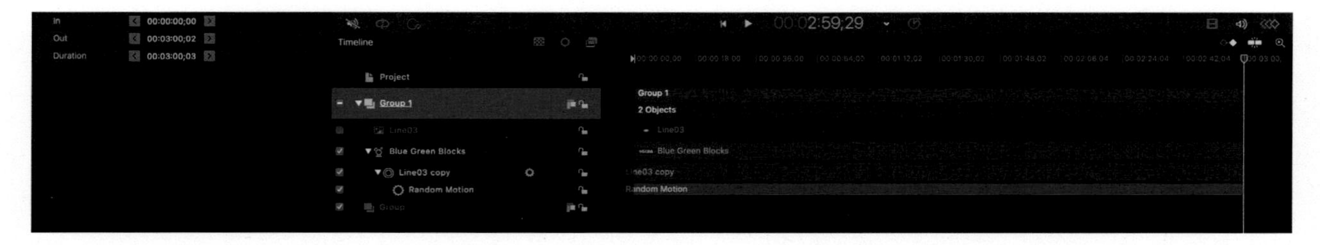

Figure 2.4 Example of a three minute timeline using Apple's Motion software. Source: Image by Author

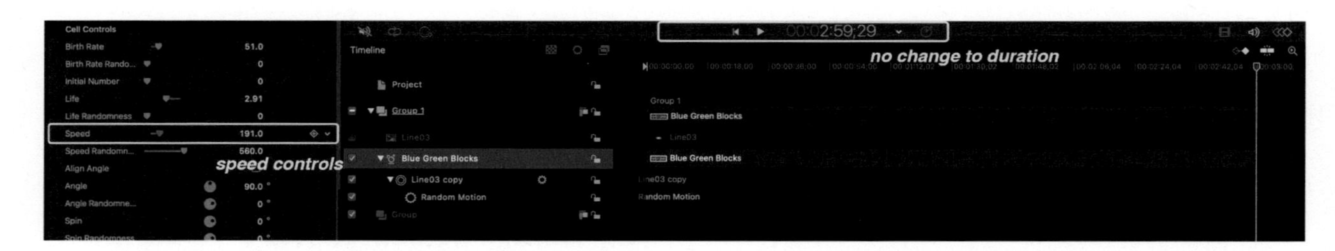

Figure 2.5 Speed adjustment using speed parameter with no change to file length. Source: Image by Author.

Figure 2.6 Keyframed events on a parameter can be attached to external data or triggers for cues. Source: Image by Author.

Traditionally, systems manage video playback as a sequence of cues. These cues often signal a change of file being played back, but depending on the playback device, can introduce color, speed, or other visual effects to the file. Real-time content files have expanded the definition of what a cue can achieve. Besides altering the look of the content with a cue, content can respond to information and events occurring in the physical performance or presentation space.

Let's take the example of a performer on stage interacting with visuals on a video screen. Every time the performer claps, the video content sparkles for 1 second. If I am the media server operator, I don't know when that clap will occur. The clap is a random moment in time and will occur some random number of times. There is an audio input device that will recognize a clap, feed the event into the media server, and trigger the content instruction set that will respond to a clap with 1 second of sparkles.

In this example, we have defined an action by a human as a trigger, a common performance cue. Our stage manager in this example expects a performer to clap and calls a cue right at the moment of the clap to start the sparkles. The result might look the same as the example previously described. However, what if we take this event off the stage and make it a gallery installation? What if we don't know when the clap will occur or how many times? What if it's more than a sparkle effect and fundamentally alters the behavior of the video imagery? What if we introduce enough variables that a computer will do a better job without human intervention?

In this case, a real-time content file will continue to deliver the defined number of video frames while responding to new commands or triggers to change what we see. Often referred to as interactive content creation, these experiences depend on real-time content production techniques.

Many types of inputs can trigger changes to video content like audio, video and infrared sensors, mechanical encoders, and optical trackers. There are many ways to respond to the information from those sensors. Before we dive deeper into the world of interactive content, one more reminder on the meaning of real in real-time content.

Real time means true to time. Real = True. The visual power of interactivity fails when "real" is no longer true to time. Input triggers that aren't processed fast enough or delayed transitions interrupt the effect of flowing through real time. The audience will either feel immersed in visuals conjured up as by magic, or a mis-timed result will feel cold and disconnected. It is up to content creators and engineers to coordinate the creative demands of the real-time content to protect a playback system's ability to remain true to time and true to the viewer's experience of the content.

Variable Triggers Combined with Variable Results

With so many variables to consider, how do we start? Ultimately, we must plan real-time content creation as a team would plan any content design. Variable definition belongs in the design process, as do their designed outcomes, even if the outcomes have some unknowns. Let's look at some of the common variables to consider when planning real-time content creation. This is not an exhaustive list, but a starting point.

Human Action:

Will a person impact the imagery?
- via their position?
- via an action they take?

Can multiple people impact the visuals at once?
- what is the processing limit?

How will the person be tracked?
- Motion Capture (MoCap), LiDAR, IR Beacon, Video camera?

Scenic Position:

Are there moving video surfaces to be tracked?
- by encoder, IR Beacon, other?
- flat or dimensional?

How is the surface tracking calibrated?
How is the projection alignment calibrated?

Camera Position:

Is the imagery dependent on camera point of view?
Are there images in front of the live action, behind the live action or both?
How many cameras are tracked?

External Data Points:

Are there physical triggers that impact imagery?
Is there data collection and processing involved in image creation?

While this may look like a lot to consider, most of the use cases for real-time video content will be defined by one or two critical factors. The design process will help clarify client goals, though not every client will deliver a complete requirements brief. Be prepared to find the precise information you need through questions, existing examples and drafts. Throughout this textbook, we will see case studies of the different ways creative teams use real-time content in entertainment and what their process entailed.

Whatever the design choices, be certain choosing a real-time content solution truly supports your project. Video content can be made with a variety of software, and we want to use the right tool for the job whether it's brand new or an old standard. Think about it like coffee. How many ways are there to brew a cup of coffee? The result is a beverage with subtle nuances in acidity, bitterness, and smoothness. Any brewing method might get you the combination of flavors you like, but I bet one stands out at making the cup of coffee you want in the time available and at the price you want to pay. Use the tool, new or old, that delivers the video production experience you need at the price that works for the budget.

I believe the strongest benefits of using real-time content creation tools are interactivity and creative responsiveness. The small file size of real-time video content also presents many advantages from delivery and distribution to file storage, but some projects produce large real-time file sizes and require complex version management. Ultimately, choose the tools that best serve the project and best use the knowledge base of the team creating and maintaining the real-time video files.

CHAPTER 3

Foundations Part 1: The Video Image

How Do Computers Display Images?

Before we break down what real-time video content is, let's first understand what makes a computer image. The fascinating and rich history of computer graphics display is beyond our current scope, nonetheless we can describe the significant technology that got us here. A reader will find many resources online to explore computer graphics history. Stories wander as often into mythology as they do fact, resulting in many disagreements over who invented various technologies and when. If you are inspired to learn more about this niche of computer history, this book's companion website contains a set of useful links to get started. Rather than a detailed history, please consider this chapter a concise origin story of the real-time graphics used in the 2020s.

> Visit rtv-book.com/chapter3 for online study resources about computer graphics history. There you will find many video examples of the technologies reviewed in this chapter.

Some readers of this textbook will read a physical book, but many will read a digital file. The digital file may exist on an e-reader, a phone, tablet, or computer screen. Regardless of the output, for digital display to work, the book file contains a set of instructions to format the text to the screen in such a way that the reader can see the words. In some formats, the digital book file may render so that the words appear in the same format as the print version of the book. For this to work consistently across devices, the individual pages must be fixed images and will load to the screen much like a series of pictures.

Alternatively, an e-reader file will contain the text and images along with instructions for dynamically rendering the formatted content, adjusting to particular device and user settings. Fonts can change size and style, line spacing can expand or contract, and lines will change length and position to adapt gracefully, depending on the device display width and height. The idea of a traditional printed "page" loses meaning, and instead we track our place in the book with location markers. The user may orient by chapter and section, but these are now organizational terms referencing the contents of the book, not its format. A code-based digitized textbook flows dynamically based on changeable reader preferences. The book renders in real time based on the instructions provided.

Computers have organized text dynamically on screens since computers had screens. The character generators responsible for drawing text onto a CRT screen gave way to VRAM that handled rasterized images and vector graphics, which were replaced by the GPUs that we depend on for graphics rendering today. Graphic image quality continues to grow at an accelerated rate to meet a demand largely driven by video games and film visual effects (VFX). Every year we see more realistic looking imagery on screen and more complex environments in games. GPU processing has become so powerful that the creative video community now competes for graphic cards with crypto currency miners and machine learning developers.

The invention of computer graphics is relatively recent. Early computers did not have video displays, but shared the results of a program with punch cards or lights that had predetermined meanings. The development of display devices could not have happened if not for the invention of the Cathode Ray Tube (CRT) in 1897 by Karl Ferdinand Braun. First used in oscilloscopes in the 1930s, CRTs reached a wider market with the invention of the television. When RCA debuted the first commercially available television at the 1939 New York World's Fair, the CRT display was both literally and figuratively behind their success. History often credits Philo Farnsworth as the inventor of television, but many other scientists and researchers made critical contributions to the development of

DOI: 10.4324/9781003206491-4

the image collection and transmission technology required to make a television function.

We find many of our earliest examples of computer graphics in the developing television industry. Prior to computers, graphics used in film and early television were slowly hand produced and physically overlaid or superimposed onto another image. Examples of computer graphics for television broadcast appeared in the late 1960s, and even then, often only as text. However basic in their infancy, quickly generating graphic information had valuable uses in many aspects of television broadcast. The quest for ever more compelling ways to display names, scores, and charts for data heavy news areas like sports, elections, and weather drove rapid growth in broadcast computer graphics solutions.

Early game development also played an important part in the evolution of motion graphics. It should come as no surprise that physicists working at Brookhaven National Laboratory made use of the lab's computer and oscilloscope to create "Tennis for Two" in the late 1950s. While many similar "experiments" were going on in labs with access to computers, many consider "Spacewar!" the first distributed video game. Developed for the DEC PDP-1 minicomputer in 1962, many researchers expanded the game and shared it around computer labs. Only ten years later, Pong (invented by Allan Alcorn, or Nolan Bushnell, or Ralph H. Baer, depending on which thread of game development you follow as the beginning of Pong) was released as the first arcade video game.

Early computer art also arose in the same era. Ben F. Laposky created abstract compositions using an oscilloscope in the early 1950s. Pioneering computer artists such as Lilian Swartz, Ken Knowlton, A. Michael Noll, and many others worked as artists at Bell Labs in the late 1960s. Swartz and Knowlton created films using algorithmically generated images, while Noll pioneered 3D animation. These artists' creative process helped inform researchers about how people interacted with computers. Other early computer artists include John Whitney, whose work in the late 1950s formed the beginnings of computer animation, and Paik Nam June who is arguably the first true video screen artist.

You will find videos of these early games and artistic works at rtv-book.com/chapter3

The success of Braun's CRT technology led to the creation of video displays that were still in common household use more than eight decades later. Other display technologies, like LCD screens, have since replaced the CRT. Similarly, we continue to use established methods of graphic imagery generation that first appeared over 60 years ago.

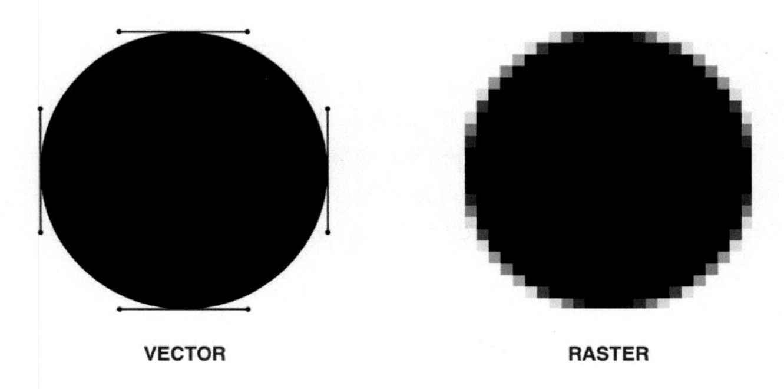

Figure 3.1 Raster image compared to a vector image. Source: Image by Author.

A graphic image on a screen is either an organized collection of pixels or the result of a mathematical equation. Bitmaps, or rasters, compose screen images out of pixels defined by their location on a grid and their color. Alternatively, we use vector graphics to draw images based on locating vertices in space and defining the mathematical relationship between two points to describe the line or curve between them (see Figure 3.1). Define enough points and you have complex 3D models.

When building real-time video content, the software tools make use of raster and vector graphics as well as models and image generating code. Real-time graphics engines depend on these types of image creation to produce the resulting video. The decades of development briefly summarized in the preceding paragraphs give only a taste of the people and work that brought us to the world of real-time content creation in the 2020s. Let's take a look at some of the image formats in use today.

What Defines a Computer Image?

To display an image on a screen, you need pixels. Individual pixels are the tiny building blocks of a display that contain color information as defined by an amount of red, green, and blue. A computer system delivers an image to a screen by organizing a series of pixels defined by location and color. An image file that is 100 pixels x 100 pixels needs to contain the data for 10,000 individual pixels. Each pixel contains 5 data points, ***x*** and ***y*** to describe location and ***r***, ***g***, and ***b*** to describe color. That is 50,000 values for a very low-resolution image. Ten million values must be managed to show a 1920 x 1080 image file. To display a video of the same resolution, a system must refresh those ten million values 30 times a second. Our playback systems must store, retrieve, and read data in these formats as fast as possible. To achieve greater speed throughout the management of an image from storage to display, we need to reduce the file size. We achieve this through the use of different image formats and compression tools.

Visit rtv-book.com/chapter3 for videos that discuss the computer image in detail.

There are multiple ways to define an image to be displayed on a computer screen, each optimized for quality, size, color or a combination. This applies to models and video files as well. To build a real-time file, you might need to reference rasterized file container types like JPG (or jpeg), TIFF, PNG, or vector graphics files like SVG and EPS, or 3D containers like OBJ, FBX or 3DS. A real-time video file might be entirely code based with instructions to generate an image or action. The software used to build a real-time content file points to these types of assets, images, models, or code in order to build the desired video file. Within the software, defined actions make these assets interact to generate new or modified imagery.

Along with static images, vectors, and models, real-time content creation software can also make use of traditionally rendered video. Video files, often compressed in .mov format, can be manipulated and reimagined within a real-time content creation tool. Similarly, we can also incorporate a live video signal into the real-time pipeline for instant manipulation of a camera signal.

What ties all these assets together are the software tools that give direction, place, time, and motion to the resulting content. Rather than combine these assets and "print" or render the result as a fixed creation, the components of real-time content remain flexible, like the text data and formatting of a digital book. The code defines their behavior and display, and can be altered during playback. For both an e-book or a real-time video, we render data flexibly based on parameters set as the file is accessed for view.

Making Real-Time Images: Generative Content

What if you want to create imagery only from code? We commonly refer to content creation that does not depend on any external assets as generative content. Generative content creation is moving imagery that is custom built from a wide range of available programming languages. We see it most often in projects that have an interactive component that defines behavior of the resulting imagery. Due to the higher time investment and skilled developers necessary to produce content in this manner, generative content has historically been the domain of permanent installations and art projects, though you can now find some of these tools in entertainment media servers.

Check out a great article from Blair Neal, describing the ecosphere of coding languages used in generative content creation and other creative technologies. Find the link at rtv-book.com/chapter3

Try searching the internet on your own for "generative art websites." Look for sites that let you interact with visuals within the web browser. You will find several good examples using only code and your mouse to create imagery. Web examples offer a quick and easy way to see how simple generative code can be, since web-based examples must load quickly. I have posted a few of my favorites on this book's companion website.

Experimenting with these interactive demos on your personal computer or phone can produce great visuals. The web examples referenced here are readable by a web browser and meet cross-browser web development standards. When used in an entertainment context, we must prepare and host the real-time content file on a dedicated computer compatible with our production environment. Generated content files need to be hosted on a hardware platform that can properly render the file into a video signal.

Reliable hosting environments for real-time content used in entertainment applications tend to be robust dedicated servers with high-end graphics cards. These environments allow for complex image creation and asset use while maintaining the required frame rate. Optimizing your code and assets will help maintain frame rates, but the demand for complex imagery often pushes the boundaries or exceeds available computing power. Make sure you have a testing plan on the production platform that will be used onsite.

Making Real-Time Images: Software and Game Engines

While real-time content creation can be entirely coded from scratch, there are software solutions that prepare executable files for real-time playback. These files are used instead of a rendered video file and maintain many of the benefits of a generative software solution. Instead of writing code to produce video imagery in real time, a content creator can employ software to simplify the production of generative content. Modules of actions are combined together with asset files, external information and other modules to create the desired visual result. We will look at specific software applications and game engines in later chapters.

Some of these software platforms are specifically designed for video content creation. They even resemble traditional video creation tools with a timeline or cue-based event structure. Most contain module or node systems as building blocks to construct the desired result. If you are familiar with traditional creative video creation tools like Adobe After Effects, these suites will provide a new approach to a familiar process.

Game engines were not built for creative video production, but have become valuable tools for real-time content creation. These platforms contain regularly used software resources to support game development. Many games rely on the same principals of a 3D world and a set of rules governing the behavior of objects and players in that world. These engines have supported the development of many successful games since their debut in the late 1990s.

How does something designed for the video game industry make its way into creative video production? Game engines became a popular previsualization tool for the film industry. Their flexibility and real-time capabilities provided a useful resource to prove shot concepts and plan VFX. As image quality improved, game engines were used on set to provide generated backgrounds to capture in camera. This practice led to their use in other entertainment design applications as a content creation platform. We will take a deeper look at game engine use in later chapters.

Check out game engines in use for film previz at rtv-book.com/chapter3

We have a variety of ways to generate imagery in real time. Video has been generated to screen as long as there have been screens attached to computers. While the current use of real-time creative video production has reached dizzying complexity, the beginnings are humble. Start simple and build your expertise.

There is no one perfect approach to deliver real-time content. In fact, multiple tools might be required to achieve your content production goals. Alternatively, your content production goals might be achievable with more than one of the available tools. Don't worry about picking the right tool to start out, just start. The knowledge you gain will have value that crosses over to other tools and application uses.

Remember this incredible fact: we've gone from playing tennis on an oscilloscope to real-time generated content production capabilities in only 60 years.

CHAPTER 4

Foundations Part 2: Code and Generative Content

In this chapter, we will focus on the development of the currently available real-time video content creation tools. Let's first consider the most common type of real-time video content, the untreated video footage captured by a camera: a live video signal. Anyone with a smart phone and internet connection can deliver real-time video content. This is the basis of a live broadcast, whether conventionally delivered to a television or over the internet to a website. Instead, this textbook focuses on the use of a computer to generate or augment video imagery in real time. Such imagery includes motion graphics, animation overlays, and any other computer created image content including content that is generated and composited with a live video signal.

When we discuss real-time video content creation, we must also consider the graphic software used to generate images. Advancements in one aspect of real-time video content creation will inform the tools in another application of real-time video. For example, a new feature in gaming content creation may power a work of video art, while advancements in web content creation might inform game content creation. Software tools often migrate beyond the goals of their original design to inspire creative use in alternative disciplines. We saw this in Chapter 3 in the discussion of game engines used for creative video production.

The live entertainment industry benefits from a host of technologies originally designed for other practices. Video games, film VFX, the internet, and more have all improved essential technologies that make image content faster to render and deliver to screen. This chapter serves as a survey of those technologies and their origins.

Generative Content

Generative video content is a subset of Generative Art, an artistic discipline encompassing work from many artistic practices, including music, writing, and a variety of visual media. Any defined process governed by a set of rules can create generative art. Let's use dance choreography as an example. We will create a new piece (or several new pieces) based on a simple set of rules:

1) Dancers use the Fibonacci Sequence to define the number of steps to take moving forward in a straight line.
2) When the number of steps taken matches the current number in the numeric sequence, turn 90 degrees to the right and continue to the next value in the sequence.
3) If a dancer runs out of room to complete the current value, they turn 90 degrees to the left and continue until completing the steps or meeting another obstacle.

By giving a group of dancers a random set of starting positions on a dance floor and these instructions, the dance company will generate a unique piece each time they perform (see Figure 4.1). This is a form of generative art. The dancers, their start positions, and the rules generate the dance piece based on the instruction set rather than following choreography determined in advance.

Generative video art advanced with the increased availability of computers in the late 1960s. Before generating video compositions direct to screens, creators used computers to control pen and microfilm plotters that generated images that became cells within film animations. Mentioned in Chapter 3, early computer art pioneers like Lillian Swartz, Ken Knowlton, and A. Michael Noll and their peers remain an inspiration to anyone creating video art with code.

> Example work of early computer art pioneers can be found in Chapter 3's companion web page at: rtv-book.com/chapter3. Check rtv-book.com/chapter4 for contemporary generative video artists.

DOI: 10.4324/9781003206491-5

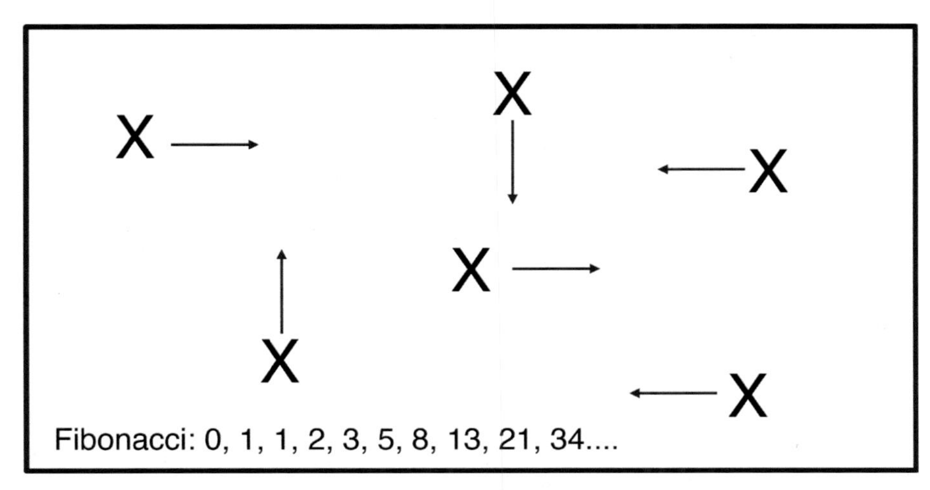

Figure 4.1 Fibonacci dance map showing starting locations and direction for performers. Source: Image by Author.

As a principle, the instruction set for creating generative video art must remain somewhat autonomous, with the artist setting the rules that govern the visual production. Writing code to generate visuals is a collaboration between the artist and the computer. The artist defines the limits of that collaboration on a spectrum from detailed fixed constraints to broad unpredictable randomness. As with our dance example, the artist's rules define boundaries within which something new can happen with each execution.

Generative video art is an excellent entry point for understanding the origins of real-time video content production. In the early days of computer art, real-time meant computer time or as fast as one could produce the result. As graphics processing advanced, the artist defined the desired time and set the visual pace of the image result. What has remained consistent is the continued collaboration between code and artist to generate visuals.

An interesting sector of generative video art is Demoscene. Arising with the popularity of home computing in the 1980s, Demoscene combines coding, motion graphics, art, and a bit of competition to create Demos. Demos are computer programs that create audiovisual displays with as few lines of code as possible. The Demoscene subculture strove to get as much power and visual complexity out of limited 1980s computer graphics environments as possible. Demoscene perfectly merges the technology of its day with creativity and competition.

Demoscene continues to this day even as the graphic computing power of desktop computers has increased exponentially. We can see the lasting influence of Demoscene on both visual design and today's real-time software applications like Notch. Demoscene altered the concept of how to share video files through packaged code rather than rendered video.

Get more insight into Demoscene history at rtv-book.com/chapter4

If the thought of using code to hand craft your own generative video art is exciting for you, you have several programming languages to choose from. While you can use any coding language for creative purposes, some are more commonly used than others for creative coding. Processing, Vvvv, Cinder, and Max MSP enable the rapid creation of visuals using coding frameworks. Some of these suites are so intuitive that they have become platforms to teach people how to code.

Check out rtv-book.com/chapter4 links to creative coding languages and tutorials

For further research, I recommend searching for #creativecoding on Instagram or elsewhere online. I regularly follow this hashtag for inspiration, not only for the art, but to explore new technologies and their visual applications.

Video Games

It will come as no surprise that the video game industry leads the development of computer graphics hardware development. The market is so large and profitable, and specific games so popular, that makers can re-release the same game with changes in storyline to take advantage of new GPU developments and sell more games.

Video games also drive improvements in real-time content creation. While waiting for the next generation of GPUs, game developers have to squeeze every cycle out of an existing line of processing power to make a game look its best. This means advances in software and image processing to improve compression, ray tracing, physics emulation, subsurface scattering, ambient occlusion and so on. For the latest topics in advanced graphics rendering, check out the papers presented at an upcoming SIGGRAPH conference.

The combination of processing power with advancements in rendering science means video games continue to look more and more realistic. Compare the various iterations of image quality across versions of a popular video game that has been around for 20 years or more. There are examples from the game Halo on this chapter's companion web page that show how it looked in 2001 versus the latest release in 2021. The image quality and game processing capabilities have advanced to the point that making a game look like a late 1980s pixelated 8 or 16 bit arcade game is a style choice rather than a limitation.

So how does video gaming inform real-time video content production? A video game is a computer program that typically runs on specialty hardware, pointing to custom creative assets to generate the world of the game. If the gamer turns the controls to the right, the game world needs to render to screen in time with the gamer's movements. If the gamer picks up a weapon, that prop needs to be rendered. Game builders cannot predict any of these actions to pre-render the visuals.

For game play to feel real, the rendered result of gamer action must happen in real time. Any lag in action or delivery of image to screen makes the game feel unnatural or even broken. This effectively becomes the processing boundary in game image quality: draw all the polygons you can in real time until it breaks the user experience of gameplay.

Games have gone from the wireframe "3D" world of 1980s Battlezone to the futuristic realism of games like Halo. There are many companies that have pioneered game engines that allow rapid use of world building tools for game development. Two of these tools have found a home in entertainment visuals: Epic's Unreal Engine and Unity.

Epic released their Unreal Engine in 1996 along with the Unreal Game. The game developers had coded the same world building tools so often that it made sense to build it as an engine for reuse. The game engine has become far more successful and across more disciplines and users than the game that inspired it. This includes its use as a real-time video content creation tool. Unity came on the scene in 2005 to support 2D and 3D game development. It has also found use outside the game industry for use in real-time content creation.

Both of these game engines currently find regular use in the production of creative video content and as backbones for Virtual Production playback. We will look at their use in more detail in later chapters. You can find links for these software tools at the companion website.

The Internet

Did you buy this book on Amazon? Or maybe at the publisher's website? The inventory of online stores is fantastically large. To maintain a website describing every item sold by these companies requires code to access databases of images, descriptions, and product details in order to generate web pages on demand. To keep a pre-made (or pre-rendered) web page for every item to be viewable on every browser platform would be nearly impossible. The web is dynamic and most web pages render on demand based on user input and design layout parameters.

Yes, most web pages render upon user request. You might be shopping for blue sweaters, size medium, with a hood and pockets made from wool. There usually isn't a page ready-made to display a hoodie matching your specific request. Instead, the web code generates the page or the relevant elements of a page to fulfill your specific request at runtime.

Delivering image content to a web page is one of many optimizations that real-time video content benefits from. Creative coders use the same libraries that fill our web pages with imagery to create artistic visual displays. Processing, a creative coding framework mentioned earlier, has a JavaScript variation called p5.js. You can try it out for yourself using an online tool like codepen.io.

Creative video production also benefits from image compression and optimization largely created to make video files as small and with the best quality as possible for delivery over the internet. We use the same image containers in our content production.

Web browsers can also display 3D models. WebGL enables the web browser to read and display 3D information and project a 2D section view of the model based on user input. Short for Web Graphics Library and developed in the late 2000s, a wide variety of creators use it for building games, modeling structures, and creating art with code. A web browser processes the WebGL code in real time to display and allow user interaction of a 3D model. Real-time viewing of 3D models in your web browser has its limits, and for WebGL, those limits can include anything from polygon count to lighting calculations relative to the quality of GPU on your computer and the speed of your internet connection.

To see example of 3D on the web, please visit rtv-book.com/chapter4

Tools to create dynamic graphic content via a web browser continue to develop. WebGPU will eventually overtake the capabilities of WebGL and become the new standard for real-time graphics rendering on the web. I am very excited about the future of 3D display on the web as a previz tool that puts a lot of power in the hands of the viewer. If the migration of software continues on the current trend, in time, we will regularly be able to access the content creation software tools for real-time video content files within a web browser.

The internet continues to overtake legacy broadcast viewing portals, like television. Browsers provide consumers a dynamic viewing environment based on delivering images and visual information as fast as possible in line with the user's momentary choices. By comparison, television remains a fixed medium. There will always be a place for our favorite movies and TV shows, but it seems inevitable that more and more interactive entertainment options will find a home on the web.

VR and AR

Real-time content creation technology also borrows from developments in VR and AR, Virtual Reality and Augmented Reality. Creators, innovators, labs, and entrepreneurs have experimented with VR/AR for decades, but the technology has yet to achieve wide adoption (see Figure 4.2). The headsets can be heavy and some still tether the wearer to a computer or console. Set up requires multiple devices within a room to locate the wearer. Physical movement can be limited to a small area if available at all. AR glasses continue to improve, but are still priced above VR headsets.

Despite the ongoing technical and interface challenges, development on these display platforms continues, providing rich tools for real-time video content. For example, methods for locating users in space provide a valuable means of tracking performers and presenters in real time. Employing a combination of infrared (IR) beacons, gyroscopes, accelerometers, and other tools, VR platforms identify a moving user and place them properly in a 3D environment. We also employed this technology to understand movement in six degrees of freedom (6DOF) allowing a user to move in X, Y, and Z space.

While the VR platform tracks a person's or other user's position, the scene or environment viewed surrounding the user must update in real time. If the user walks forward, the scene must reflect that motion. When the user turns their head to the right or tilts their head, the scene must reflect that change in orientation. If the user sits down, again the scene must move naturally, in real time. Any delay will make the scene feel artificial (or worse, cause headaches and nausea for the user). In fact, most VR platforms use prediction engines to anticipate

Figure 4.2 Left: VR headset and user Right:AR Headset and user. Source: Image Credit: iStock.com/max-kegfire and iStock.com/Georgijevic.

Figure 4.3 Shot of a green screen studio with a tracked camera and live feed of video composite with background. Source: Image Credit: iStock.com/gorodenkoff and iStock.com/mppix and author.

the user's movement, so the most probable frames queue in advance for faster delivery to their display. As these tools develop, VR headsets will track eye position and use multi-focal distance screens to make the visual environment more and more responsive to user action.

The science of position tracking to make the VR environment feel as realistic as possible to a user within that environment is now used to track camera positions relative to a performer or other moving target. Tools that track a human in space can monitor the position of a camera and calculate the camera perspective of a 3D environment in real time. Creators can use the result to populate video screens or composite a live image to a green screen instantly corrected for the camera's point of view (see Figure 4.3).

Variations on those positioning tools for VR also exist in a modern cell phone. Tiny gyroscopes determine if you hold your phone vertically or horizontally, or use it as a steering wheel to drive a game car around a looping raceway. In one package, you can define your location in space, generate graphics, view real-time composited images, and share the result over the internet.

While phone gaming tools drive the technology, we can use the same tools to create AR graphics for concerts. The possibilities continue to grow for creative applications on smartphones. Now that the newest phone models also contain Light Detection and Ranging (LiDAR), generated graphics can "sense" the 3D world around them. Phone LiDAR systems detect the time it takes surfaces in the environment to reflect pulses of light invisible to the human eye. The result is a depth map or a rudimentary 3D model of the space in front of the LiDAR sensor.

We have an incredible number of ways to generate content in real time in 2022. From custom code to game engines, from the internet to the promise of VR, an age-old challenge remains: finding the right content development environment that best serves the project. No single toolset perfectly suits all projects, and no project succeeds purely based on the use of any particular toolset.

Let's look at a case study from a 2010s project that used openFrameworks to see how Blair Neal approached the challenge of real-time interactive content creation.

Case Study – Blair Neal – Trace Your Road

Creating a life-sized racing video game is no small task. *Trace Your Road* was an experiential event for Lexus brought to life

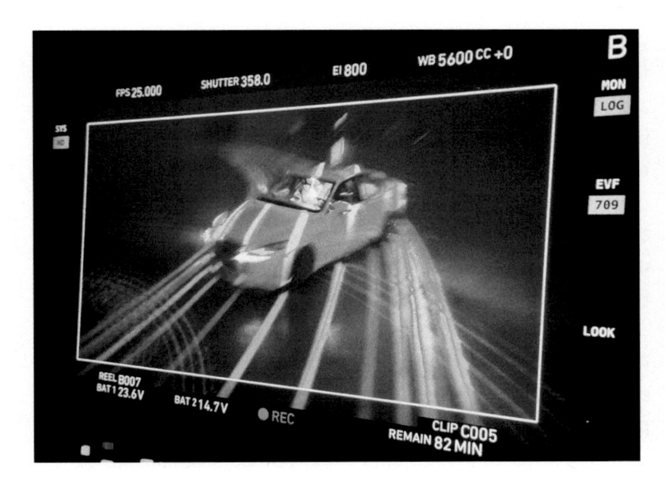

Figure 4.4 Car traveling along an interactive projected path.
Still from Lexus *Trace Your Road* video.

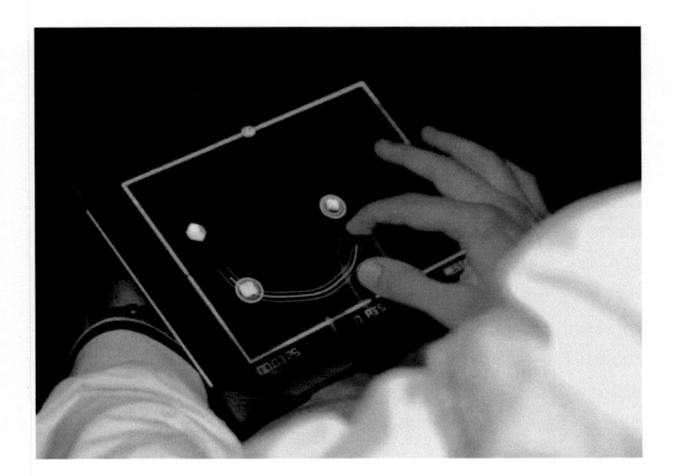

Figure 4.5 User interface from custom software.
Still from Lexus *Trace Your Road* video.

Figure 4.6 Driver and user in the car.
Still from Lexus *Trace Your Road* video.

in 2013 by Fake Love in an airplane hangar on an Italian military base. Lexus selected a small group of Formula 1 racing enthusiasts to take part in a thrilling experiment using real-time virtualization technology. Figure 4.4, Figure 4.5, and Figure 4.6 detail some highlights of the experience.

Visit http://rtv-book.com/chapter4 for supporting video and images for this case study.

Participants took the passenger's seat next to F1 superstar Jarno Trull, in a race to hit checkpoints as quickly as possible. Participants geared up with an iPad and a helmet then strapped into an F1 car. The car moved into position and the countdown began. When participants drew a line on their iPad, the same digital line would be projected in real time in front of the car on a 220 ft x 120 ft projection surface. Trulli would punch the gas and do his best to stay on the traced line. The participant's goal was to hit five randomly positioned virtual checkpoints (both projected and visible on the iPad) as quickly as possible. They would trace swirls, straight lines, and sharp corners. Trulli went 30–40mph and drifted like crazy in the hangar in order to hit each target as quickly as possible. Most participants finished in less than a minute. A leaderboard in the green room told them how well they ranked after each of the subsequent rounds.

The event was a huge success. Lexus' resulting television commercial and ad campaign received excited media coverage and millions of views on YouTube. Many technological hurdles tilled the road to this singular achievement. Let's dive into the details a bit more.

The project itself moved from concept to completion in a little less than eight weeks. The eight-week timeline included creative concepting, gameplay design, asset creation, hardware planning, custom software development, and onsite work. Very little time remained for testing and iterating on any proposed solution. For many things, the team had only one shot to get it right. As a commercial project, the team did not feel the pressure of building an installation that would need to live for a year or more, but their racing game had to work in real time to deliver a genuinely great experience.

The entire event ran on four applications programmed using openFrameworks, an open source creative coding framework written in C++: the participant app on the tablet, the projection rendering application running on Mac Minis, a master renderer or controller, and the computer vision application running on the Mac Pro used to track the car's location on the projection surface using a feed from an overhead camera. These applications communicated fairly simple messages over OSC. The master rendering app ran on one Mac Mini and handled most of the logic. The rendering clients (as opposed to the master) ran on four Mac Minis, each rendering a portion of the overall canvas. These canvas portions were cut up and sent to one of 12 20,000 lumen projectors mounted in the rigging about 45 feet off the ground. Projection at that scale ran a risk of washing out because of room lighting and the dark concrete floor. While the final product looked dimmer than hoped, it still worked with the high contrast visuals and proper camera lenses.

The team chose openFrameworks as the real-time framework, primarily because of familiarity and availability of developers. At the time of this project, other sophisticated options did not exist or were not yet mature enough to support something on this scale in six weeks. openFrameworks could render real-time graphics, do computer vision tracking, communicate over the network, and use a similar codebase across desktop and mobile devices. We also had some budget challenges that limited the ability to use high-powered computers and other things, hence the use of Mac Minis.

While most of the production went off well, we faced several challenges when we arrived onsite. These challenges primarily came from a lack of testing time with real world conditions, but also limited access to equipment. We also confronted some evergreen challenges like stable Wi-Fi connectivity to the tablet in the car – there were eight wireless GoPro cameras crammed into the car that made the signal weaker than tested prior to the shoot day. We had challenges with getting our Matrox Triplehead2Go to reliably connect to our Mac Minis and had to get backup units quickly (not an easy task in rural Italy). See Figure 4.7 for a wiring diagram of the project.

The biggest challenge we faced was related to tracking the car on the projection surface. When developing our initial assumptions, we considered a number of options:

- LiDAR – too expensive and too slow for a fast car, and specialized enough that we wouldn't have enough development time to integrate
- GPS – too slow and inaccurate indoors
- Depth cameras – not enough reliable range when mounted on the ceiling
- Multiple cameras – too expensive and logistically complicated to test and align multiple cameras, especially back then and with only a week onsite

We ultimately settled on a single camera with an extreme fisheye (160°) lens that mounted on center over the projection area. The camera had a filter on it to only see infrared (IR) light so that the computer vision tracking algorithm wouldn't confuse the car with the projected images. In order to make the car easy to track, we put a couple of high powered IR lights in the rear window pointing up. In an ideal world, we would have mounted these lights on the trunk to make them visible at all times, but we could not alter the exterior of the car.

We tested the lens and tracking algorithms in a small studio space with a 12ft ceiling and a remote control car a few weeks prior to going onsite. The test was successful and seemed to show that this solution would work, but we still considered it a bit risky. Unfortunately, when we got onsite and positioned the camera, we quickly realized that the extreme fisheye necessary to capture the entire field created some problems. Tracking the car worked, but the curvature of the fisheye caused the tracking to show a curved line when the car traveled dead straight. Due to onsite production delays, we discovered this only 48–36 hours before the live date. We had very little time (or sanity) left to implement a camera dewarping algorithm or other solution into the system and test it appropriately. We quickly shifted to Plan B.

Since the automated tracking wouldn't work, we instead relied on knowing when the car hit the checkpoints. The solution was: manual tracking with a wizard behind the curtain. We altered the tracking application to allow for manual input via a mouse, rather than the computer vision algorithms. The modular design of our system allowed for this piece to work independently to send positional X/Y coordinates to the master rendering application. An observer would sit in the ceiling and see a very faint blue circle on the projection indicating their mouse position. When the car moved, the observer would do their best to keep the circle around the car, which caused things like checkpoints and finish lines to trigger as expected. With about 20 hours to go, we tested the solution. It worked without any major hitches. Countless runs were completed while the film crew captured things from every angle.

If this project were to come up today, I think there would be some very different approaches and considerations for both software and hardware. For starters, I think we would need to significantly extend the development timeline to allow for testing and making things more robust. Most of the issues onsite related to an inability to test with real world conditions and even though the team made their best effort to simulate those conditions, there were still (and will always be) unforeseen issues. Making a risk assessment plan prior to arriving onsite would be critical for identifying potential points

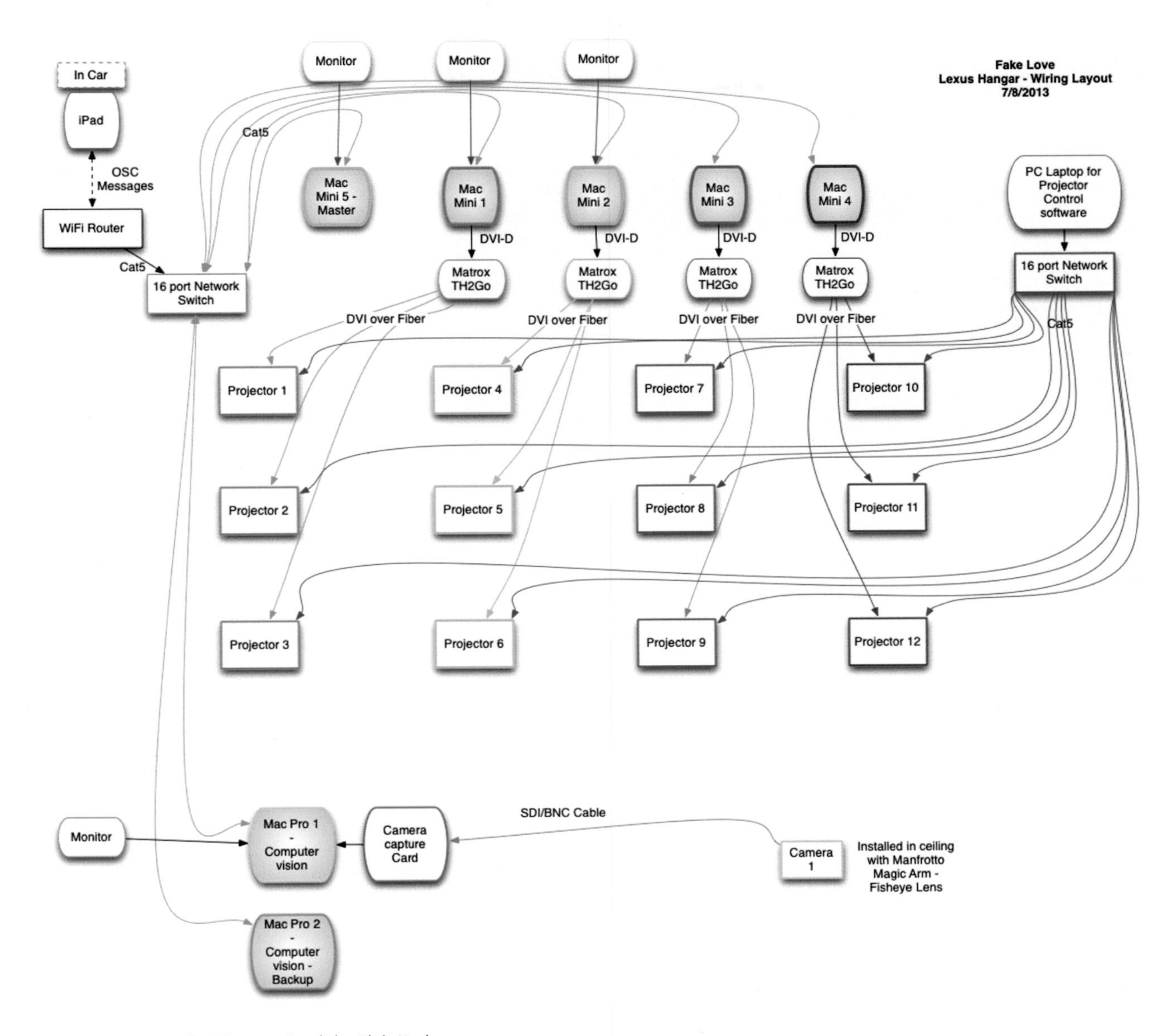

Figure 4.7 System Wire Diagram. Graph by Blair Neal.

of failure and alternative solutions. Making those plans ahead of time with a level head will often get you further than having to decide during a stressful onsite.

While openFrameworks would still be a valid choice for software development, I think TouchDesigner is probably the more obvious choice for such a large canvas and synchronizing multiple displays. A game engine like Unreal or Unity could also work, but we would need to consider how to split up the canvas for projection. The canvas at the time was nearly 4k resolution, so rendering all of that on one computer makes a lot more sense than in 2013. Once you have that digital canvas, you could use video hardware or another system like disguise or Watchout to split up and distribute the image to the different projectors or do some other crazy multiple graphics card setup to get 12 outputs. The tablet application would still need to run as a native app, but I think a web application would also provide a valid approach these days. See Figure 4.8 for a screenshot of the openFrameworks project file.

In terms of hardware changes, while projection was a little dim, there isn't really a new solution that would work in the same way. I think adding more projectors or maybe painting the floor or something could have helped improve contrast. As for a car tracking solution, I think the IR camera path is still a valid one. I would want more time to test and would add

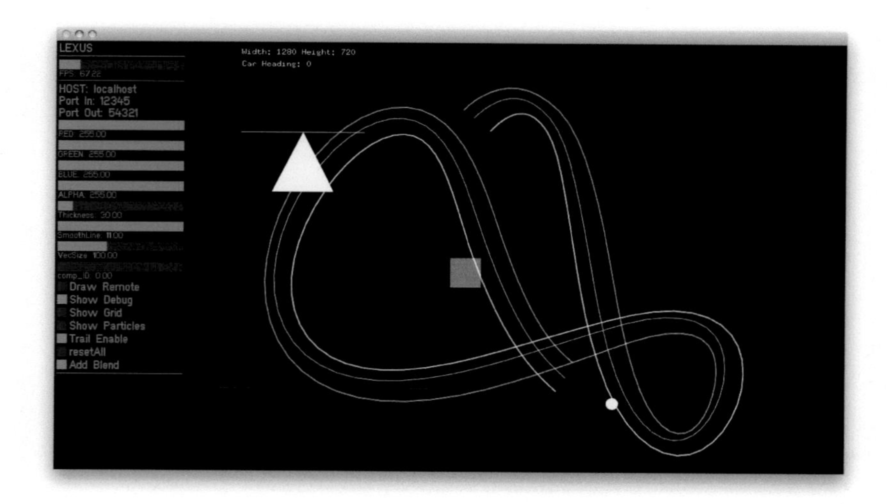

Figure 4.8 Software Screenshot. Software by Blair Neal.

in a lot of extra features to handle things like correcting the fisheye curve and maybe some machine learning algorithms for better detecting the shape of a car rather than only an IR light. I think it would be more straightforward to simulate some of this camera setup in a previz environment these days. Back then, we worked as quickly as possible with some 2D tools. Additionally, doing things like processing and aligning multiple cameras would be easier now with added computing power and more input options like NDI. LiDAR would be my next best selection because the tools have become more accessible and easier to implement. I would also plan to have manual tracking via an observer as a backup solution rather than a hasty onsite decision.

At the end of the day, even with the technological tools changing every year, things like planning, testing, and having alternative options never go out of style. There are always multiple solutions for any given problem, but making choices that allow flexibility for creative exploration while providing robustness for technological success always yields great results.

By Blair Neal on a project for Fake Love
https://ablairneal.com/

Blair is a technologist and artist who helps artists, agencies and brands creatively and purposefully apply technology to human-centered experiences

Project: Trace Your Road
Location: Rome, Italy
Year: 2013
Team: Saatchi and Saatchi, Logan, and Fake Love
Client: Lexus

CHAPTER 5

Foundations Part 3: Broadcast

We cannot discuss the history of video and projection graphics for entertainment without also looking at the technical evolution of graphics and compositing in live broadcast television production. Prior chapters outlined some of the development of real-time graphics for broadcast applications. They began with simple character generators, advanced into sports and weather displays, and recently matured into Augmented Reality and scenic Background Replacement. What follows is a brief review of this history and the origins of some of today's most popular tools.

Is it film or is it television? We think of film as a two(ish) hour length storytelling presentation and television as a multi-episode presentation of many stories around a single theme. We also use "film" and "television" to differentiate production styles. Film tends to be a single camera point of view, while television employs multiple camera perspectives. Film style production is used on many television series and multiple cameras may record concurrently on a film shoot. We currently do not have a clear terminology to delineate shooting styles, and the workflows can be vastly different between production types. This chapter covers multi-camera live broadcast style production for television.

Studio Image Compositing

We can categorize television broadcast content into information, entertainment, or a hybrid. In this classification, information includes news, documentaries, and educational material, while entertainment encompasses narratives, game shows, live events (like rock concerts and awards shows), reality shows, and sports. As television continues its migration to an asynchronous internet transmission service (i.e. streaming), rather than a scheduled or synchronous broadcast service, I believe these categories continue to hold true.

For example, news programs provide an information delivery service, while your favorite episodic series delivers an entertainment service. We might consider sporting events simultaneously as entertainment and information broadcasts about the sport. While watching the sporting event, the viewer receives statistics about the game. Our particular interest is how the broadcast workflow combines live camera footage with informational displays in real time.

Can you imagine a newscaster reading the news without bullet points of critical information displayed on the bottom or side of the screen? Or a football game with no colored marker to indicate the current scrimmage line? Or a character in a TV series sending a text message without some display of information showing message contents? These examples all use camera footage, live or prerecorded, combined with computer generated imagery to enhance information. Two video sources are composited together to create the final image, often in real time.

First, let's break down what happens in a television news studio. Imagine a newscaster sitting at a desk, staring at a camera, reading a script from a teleprompter directly in front of the camera lens. See Figure 5.1 for a mockup.

The orientation of the camera to the news anchor is highly controlled. Desk, newscaster, and background positions are structured in the image frame. Planned negative space around the shot allows room for text boxes, text based news crawls, image references, and talking heads to appear alongside the newscaster. Figure 5.2 outlines the composition of the shot.

These types of graphics are pre-produced and the signal compositing happens live. The camera signal feeds into a mixer along with the signal of the graphic overlay. Depending on the technology used, the graphic image signal comes to the mixer along with a signal describing the usable information as a matte layer, sometimes referred to as a cut and fill. The signal compositing technology in the mixer cannot correctly combine

DOI: 10.4324/9781003206491-6

Figure 5.1 Typical news studio camera shot of presenter at desk with graphics overlaid onto image. Source: Image Credit: iStock.com/Tetiana Musiyaka and author.

Figure 5.2 News studio example with highlighted areas showing protected regions for camera framing. Source: Image Credit: iStock.com/Tetiana Musiyaka and author.

two video signals without information defining which parts of the graphic overlay signal to use and which parts to discard. Figure 5.3 describes this signal compositing process.

In the 1960s, character generators were developed to create and display text overlay information onto a signal in real time. Using a luminance key, broadcasters superimposed white text on a black background over the video signal. We still use this type of keying in graphics software, with the black portion of the signal assigned to transparency and the white text appears opaque.

Other types of keying terms that you will continue to hear today include, superimpose, cut and fill, and matte, even if the underlying process at work has changed. Terminology is slower to adapt, often referencing the origin of the action described. For example, "Roll tape!" is still often shouted across the broadcast operations room, even though that tape was replaced with hard drives years ago.

While character generators have grown into complex systems capable of handling motion graphics and 3D assets, the general concept remains the same. We predefine some portion of the live video signal as available for graphics use while controlling camera and newscaster positions to keep the graphics area clear. Humans at the camera controls must properly align the shot for the composite to work. These tools are the stepping stones to broadcast AR, which will be discussed further in Chapter 9.

Figure 5.3 News studio example in a layer view outlining the cut and fill process. Source: Image Credit: iStock.com/Tetiana Musiyaka and author.

Next, let's consider another aspect of real-time television image compositing: Background Replacement. At first the term "Background Replacement" may seem unfamiliar, but if I call it "green screen" or "blue screen," I bet it is quite familiar. Video conferencing and social media tools use the term "green screen" as a stand-in for Background Replacement. Software solutions like these separate your image in the foreground, from the home office mess in the background using only the data captured by your laptop or phone camera. When we capture video of physical objects and people in front of a monochromatic, well lit background, the physical process of keying out that background color is called Chroma Key. Any color will work, but primary green is the most common key color (see Figure 5.4).

When used as a background, a broadcast mixer is employed to ignore or remove the key color from the original signal. The mixer can then composite the image over another video signal. Using state of the art modern systems, the results can look quite impressive and occur in real time.

> For videos on green screen history, visit rtv-book.com/chapter5 for further insights

Historically, the concept of green screen originated in film production and did not happen in anything close to real time. Key color removal and compositing required lengthy and often complex post-production processes. The celluloid film used in pre-video capture filmmaking relied on a light and chemical based process to key out color. This required clever use of prisms and film development techniques to make a visual effect look plausible to the viewer.

Before digital television and the computer aided removal of specific colors, an analog video signal keyer could identify colors via signal analysis. While this process wasn't perfect, it was effective. Initially, studios used blue backgrounds, borrowing the standards from film, but eventually green was the most common color choice. Broadcast cameras of the era were more sensitive to green, the color was easier to light, and more often contrasted with costume and prop colors.

Green screen became the standard for Background Replacement in weather reports and late night talk show comic bits. These uses revealed the limitations of early real-time compositing. Sometimes the lighting between recorded background and live foreground didn't match, the compositing was rudimentary, or the background itself was not dynamic. We'll call this the era of the video backdrop. While the background can look complex, the perspective is fixed relative to the camera as if it was a static wall. Today, we employ an interactive production environment of camera responsive digital backgrounds, foregrounds, and image augmentation. To take this

Figure 5.4 New version of news studio with green screen layer outline. Source: Image Credit: iStock.com/Tetiana Musiyaka and iStock.com/monsitj and author.

next step to dynamically respond to the camera's point of view, we need to know where the camera is looking.

Camera POV

Like a painted backdrop, a scenic video backdrop is not responsive to camera position. Certainly you can have a sophisticated video as your background, and that might feel dynamic and engaging. But like any treated scenic wall, it does not have depth and will appear flat as the camera changes angle. Depth can be added of course, using forced perspective in the image to trick the eye into seeing a flat surface as dimensional. This trick breaks when the image is viewed off angle, including on a video surface, unless the imagery can respond to camera POV and correct the depth perspective in real time.

Consider this example: we have a live shot of a performer about 6 feet in front of a tree. Imagine that tree 20 feet in front of a cabin, which is itself in front of a mountain. As a painted backdrop, when the camera moves, the components of this image will remain static relative to one another, exposing the lack of depth. We do not improve the sense of depth if we stage the performer in front of a green screen and use a pre-rendered video of the same environment as a background. Look at this very basic comparison in depth shift shown in Figure 5.5, versus the camera shot of these elements when layered. With a camera move, the layers can simulate what would happen in nature to fake the experience of depth. Notice the subtle differences between where the trees, cabin, and mountain peaks line up along the x axis in the reference images.

Camera Point of View (POV) is critical to understanding the depth relationships between a digital background and physical foreground elements. When the shot changes position, the relationship between these elements also changes within the frame, altering the viewers' sense of depth. If the physical foreground changes relative to the digital background in a moving camera shot, but elements within the digital background remain static, the background looks flat.

For the example in Figure 5.5, what if we found a location in the woods that matched this drawing and recorded video of a camera pan? We would record examples of the camera move across the tree, cabin, and mountain so that it exactly matched the camera pan we would make in a green screen studio. We can send a crew out for a location shoot with a robotic camera and a pre-programmed sequence for the camera move. We use that same pre-programmed

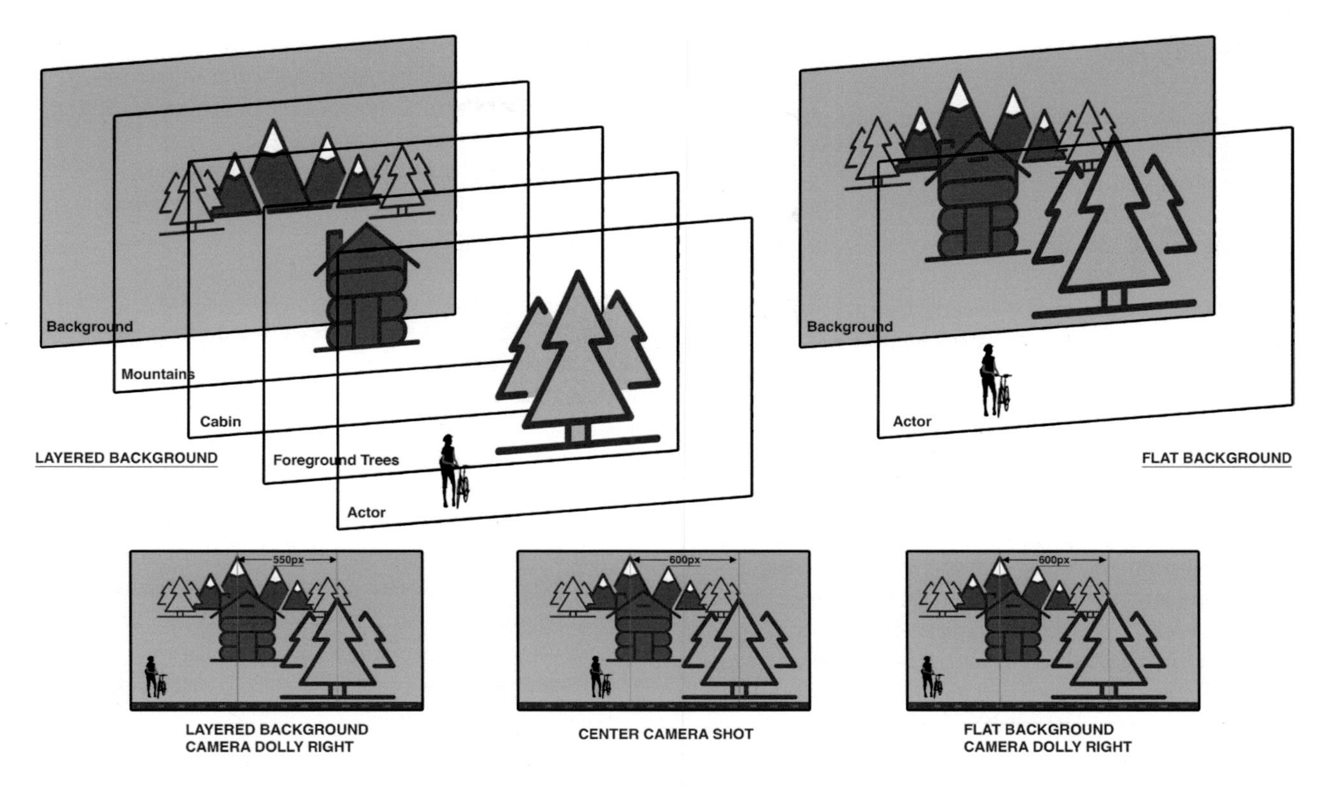

Figure 5.5 Visual example of depth in a scenic background and the result of a shift in POV. Source: Image Credit: iStock.com/TongSur and iStock.com/A-Digit and author.

sequence in the studio. We then composite the two matched shots so that we maintain a natural feeling relationship between the studio footage and the site footage. The depth relationship between all the elements moves as if we had shot the performer on location.

Better still, what if we didn't need to pre-program the camera movement or use a robot at all? Instead, we could collect the data from the camera in the studio as it is moving and recreate the same motion in a computer, digitally generating any background we need within the content creation tool. As long as we understand the lens and camera position, we can use a simulated camera to generate imagery from the same POV and composite the result with the footage. Let's consider the development of two of the technologies we need to realize this solution: camera tracking and background screens.

Sports

Sports has exploited the relationship between camera POV and graphics overlays going back to the 1980s and ice hockey. I highly recommend watching a video produced by vox.com called, "The NFL's virtual first-down line, explained" (www.vox.com/2016/2/6/10919538/nfl-yellow-first-down-line-espn) which is also posted on the companion website for this textbook.

In a football stadium, if the football location, field and the camera positions are located in 3D, and you receive real-time telemetry data of the camera POV, you can overlay real-time generated graphics onto the field surface, like a 1st down line. As the camera moves, the generated line stays in the proper location relative to the football field.

These sports graphic overlays have come a long way in the last 20 years, covering every imaginable data point a viewer might want in any type of sporting event. What country is that swimmer from in lane 5? Use a graphic overlay in the pool lanes. Need statistics on a WNBA player? Display a graphic following her down the court. Have a camera crane panning during your commentator presentation? Place informational 3D graphics hanging in space that respond to camera position.

Sports information visuals have developed to the point of incorporating camera POV to create virtual scenic elements that exist in the same "space" of the camera's field of view. Broadcasters use virtual scenery in combination with advanced camera tracking tools to create Broadcast Augmented Reality.

Broadcast AR

Broadcast Augmented Reality, or AR, is the compositing of camera sensitive, computer generated images in the foreground of a live camera shot.

Four things make Broadcast AR possible:

- Duplication of the camera telemetry from the real world into a virtual computer generated world in real time
- The ability to accurately represent the physical world of the shooting space in a 3D model
- Generation of 3D foreground assets in real time
- Composite of the result with minimal (sub-second) delays

For example, how might we make a set burst into flames for the finale of a live performance? For an indoor event, we cannot always use live pyrotechnics and can instead superimpose the flames onto a set. All we need are accurate 3D models of the set and position data from our desired tracked cameras in real time. Each camera signal will be processed through a separate computer to generate fire to composite onto the camera shot. The computers receive a camera signal and the corresponding camera telemetry. They also host the model information necessary to produce fire in the right location of the image frame. Finally, our system outputs the composite to the technical director for switching between shots on the live broadcast feed (see Figure 5.6).

Not only can we augment a set with position sensitive graphics, we can track individual performers with animated characters or effects by including their position data. We can add new scenery to expand the set beyond the boundaries of the venue. Combined with the current power of various Background Replacement tools, we can create an all Virtual Production environment, surrounding a performer in generated imagery, all processed in real time.

Background Replacement

As AR tools matured and video screens dominated more of the scenic space, the next iteration of digital stage development came into existence. AR camera telemetry data was applied to video content destined for the background of a camera shot. Live generated content for use in the background of a scene responds to camera POV and appears to have natural depth. The result is camera position sensitive content intended for physical video screens and green screen compositing.

The same requirements for use in Broadcast AR are also used for real-time green screen compositing:

- Duplication of the camera telemetry from the real world into a virtual computer generated world in real time
- The ability to accurately represent the physical world of the shooting space in a 3D model
- Generation of 3D *background* assets in real time
- Composite of the result with minimal (sub-second) delays

When an LED screen is your background instead of a green screen, the last step is no longer completed with a green key composite process. Instead, the generated background is displayed live onto the video screen.

Background Replacement with LED Screen:

- Duplication of the camera telemetry from the real world into a virtual computer generated world in real time
- The ability to accurately represent the physical world of the shooting space in a 3D model
- Generation of 3D background assets in real time
- Process imagery for background shot to video output signals
- Deliver video signals to video output screen

Therefore, we can define Background Replacement as the real-time generation of camera telemetry responsive, computer generated images in the background of a live camera shot that is either key composited with the camera capture or live captured in the camera background.

Let's briefly weigh the pros and cons of these Background Replacement workflows. Green screen production environments remain incredibly flexible, but they have some limitations. When we key out green from an image, the entire background becomes available to design and expand. However, this can require significant post production processing of the captured image for it to look like part of the scene. Green reflections on props, scenery, and actors may need to be removed. Lighting and natural reflections on actors and props may need to be added back in. Real-time green screen Background Replacement continues to improve in quality and is more affordable than other real-time Background Replacement options, making it worth considering for some projects.

LED and projection screens have become very popular in the last decade for real-time Background Replacement. A video background screen provides natural reflections that a camera can capture on faces, costumes, and physical scenery. Performers see and react to their surroundings more naturally than green screens. A studio with a large LED surround is called a "Volume" or "LED stage." Some studios include an LED floor or ceiling and

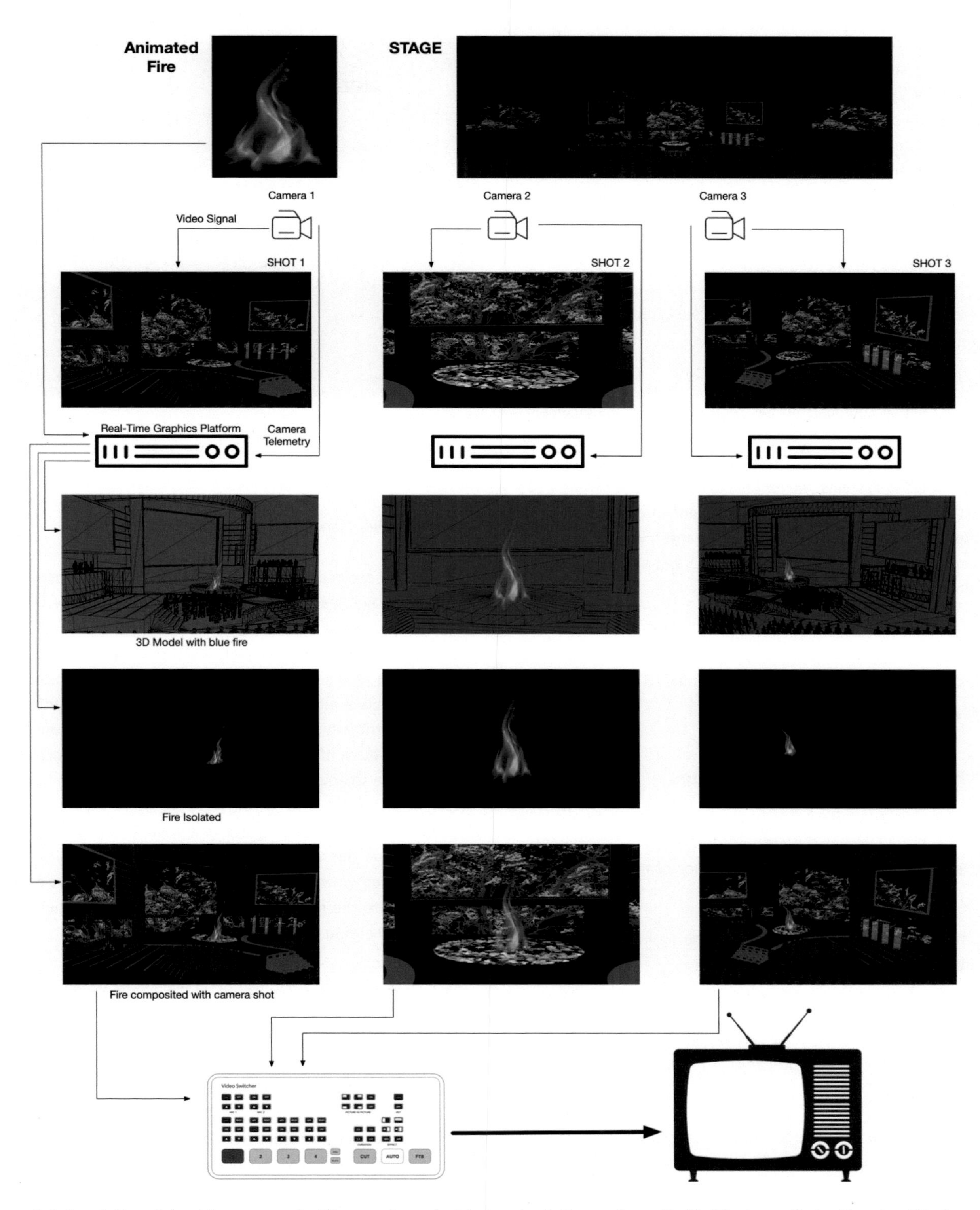

Figure 5.6 Description of signal from camera to AR processing output to broadcast. Source: Image Credit: iStock.com/Vadym Yesaulov, iStock.com/dinosoftlabs, iStock.com/fad1986, iStock.com/Ruangdej Chulert, iStock.com/seamartini and author.

shape the vertical wall areas as two or three walls or a large curve. There are some disadvantages to physical screens and real-time background content creation. Screen size can be quite large and the process expensive. The knowledge base required to deliver a successful LED stage varies significantly depending on whether you are working in film or live broadcast.

We will explore green screen and LED screen options in more detail in Chapter 10.

See more about video screen digital backgrounds at rtv-book.com/chapter5

Broadcast signal compositing has seen incredible advancements in its 60 years of use as a real-time compositing solution, from simple character generators to the latest spatially responsive camera tracking systems. We will explore green screen and LED screen options for Background Replacement in more detail in Chapter 10. For our case study in this chapter, we will focus on a creative use of camera position data and a small, live captured scenic projection screen.

Case Study – Bluman Associates – Dave at The BRITS 2020

The Project and Use of Real-Time

In 2020, music artist Dave was booked to perform his politically charged song "Black" at the 40th edition of the BRITs Awards for live broadcast to a world-wide audience of millions.

Visit rtv-book.com/chapter5 to see the performance video and learn more about the process.

Dave's creative directors for the event were the very experienced and innovative production designers Bronski and Amber Rimell from Tawbox.

Realizing that four of the other eight artists performing at the 02 Arena would also have pianos, Bronski and Amber sat down to design a performance that would "stand out whilst remaining true to Dave's amazing poetry." The pair didn't like the image of a "grand" piano, so decided on the more humble and meaningful idea of Dave sitting down at what would look like a table. The symbolism of this concept further increased when Dave decided his white producer, Fraser T Smith, would join him

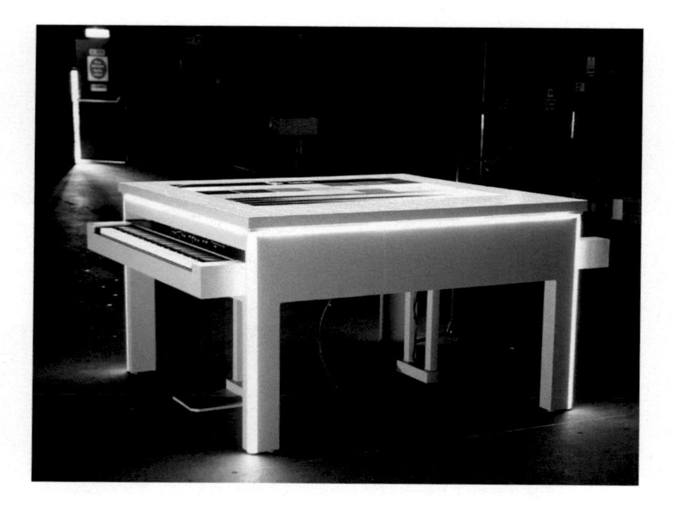

Figure 5.7 Piano Detail in the Rehearsal Studio. Photo Credit: Pod Bluman.

to play opposite on a second piano. Our "simple" table would encompass two pianos (see Figure 5.7).

Next, Tawbox decided they wanted to project powerful images inspired by Dave's passionate lyrics onto the "piano-table." Initially, these would be 2D images of newspaper headlines, flags, and words which would accompany the first verse of "Black." The images would then switch to 3D real-time projections, including prison bars, snake pits, and photos of the London terror attack hero Jack Merrit, which would play over the rest of the song's verses about black history and oppression. Everything would be shot from one steadicam and rendered in real time.

We aimed to create an immersive and dramatic merging of the virtual and the real, with the 3D images dissolving and tunneling deep down into the piano, giving the viewers a sense of peering into the real experiences in Dave's lyrics. Tawbox would use emerging XR (Extended Reality) technology.

"This kind of whole 3D world had never been done before," explains Bronski. "It ended up being a world first because the projection changed how it had to happen. We didn't want to manipulate GFX to work with LED screens – we wanted it to work with projection because we felt that would look a lot more organic."

Bluman Associates and Stout Studios were brought on board for the content and technical delivery of the project. "Initially, I wasn't 100 percent sure it could be done," admits Pod Bluman, Director of Bluman Associates, who nonetheless relished the opportunity to collaborate on such an ambitious production. "Lots of different elements needed to be put together to make one very special creative and technical execution." See Figure 5.8 for insight into the team at work.

Figure 5.8 Preparing the tools that made this projection display possible. Photo Credit: Christian Dickens.

The Challenges

Using brand new XR technology in a live broadcast was extremely challenging, but combining it with projection and a steadicam on a very busy stage meant that executing the project would be extremely demanding.

Successful execution required convergence of three state-of-the-art technologies never before combined.

1) Real-time motion 3D graphics
2) Camera tracking technology
3) Media server 3D world alignment technology and running a show in real time

Mo-Sys was chosen for camera tracking, which proved essential as their offices were based close to the O2 venue.

The disguise platform played a key role in the development of the workflow we needed to create. It was especially valuable for its ability to previz in a true 3D environment, and its unique synergy with the team's preferred real-time content tool, Notch. The Stout Studio team worked closely with disguise to develop a custom world alignment workflow just for the Dave performance, based on the XR workflows that were still in the process of beta testing. At the time, disguise was the only media solution that could achieve these goals. "It meant that everyone had to go the extra mile – and it was not an out the box solution until we ultimately proofed it on national TV," says Lewis Kyle White, Creative Director of Stout Studios and Notch artist.

Hundred percent alignment was required between the virtual worlds in disguise and Mo-Sys, and the real world. To this end, disguise provided a brand new calibration technique; a completely new application for disguise xr software was evolved.

Another challenging aspect was the need for a dedicated steadicam operator from the very start of rehearsals to the finish. "The steadicam operator was basically a performer in this show, who needed tight choreographing," says Pod Bluman. "He needed to be utterly committed to us – and Dave's management needed to be committed to paying quite a big bill for him! We were quite tempted not to let him eat for days before the show in case of food poisoning!"

Another huge challenge was using a tracking system designed for studios – not live events. "Normally trackers go up in the ceiling, which wasn't possible at the O2. There was a lot of stuff in the ceiling and it was a long way away so we opted to use the stage floor instead," says Bluman. See Figure 5.9 for behind-the-scenes on stage testing camera tracking.

"There was an unbelievable amount of 3D tracking markers on the floor, which fortunately you can't see because they were black gloss, just like the stage. Billie Eilish, Lizzo, they were all walking all over our tracking markers the entire time they were on stage," says Bronski.

The project required various proofs of concept and substantial amounts of rehearsal right up to the eleventh hour – including a full eight hours on the stage on the day as well – previously unheard of at the BRITs.

The technical complexities, lack of precedence, and the fact that the performance was not pre-rendered and was part of a global television broadcast made many people very nervous. Dave's management, the BRITs producers, the show director, and the broadcaster all needed reassurance that the technology was sufficiently robust, that rehearsals had been diligent, and that backup plans had been put in place.

Figure 5.9 Camera calibration on stage at The BRITs. Source: Image Credit: Pod Bluman.

Another considerable element that the organizers needed to agree to was that the creative team had made the bold decision to focus not on the audience in the O2, but on the millions watching on their televisions. The audience essentially watched on two screens what the television audience was watching at home, rather than what was happening in front of them on the stage.

We should state here that Tawbox, Stout Studios, and Bluman Associates' reputation, experience, and working relationships previously forged with those in charge without doubt were fundamental in paving the way to a green light for a radically new kind of performance for the BRITs.

The Tests, Research or Development to Provide the Solutions

Lewis Kyle-White was charged with the 3D Creation and Design which would involve establishing a very complicated workflow using Cinema 4D, Notch, Mo-Sys, disguise, and After Effects. This workflow would be totally new to Stout Studio and the Bluman team and indeed the industry in 2020, and needed to take many things into account – especially the flexibility to make last-minute changes wanted by the creative team.

For Lewis, it was difficult to know where to begin.

> *I jumped in to disguise and Notch, so we were at the core using a media server playing with things at the start. We did some early tests having a camera fly around a simple cube which acted as a proxy table. What was essentially happening was that the camera was passing its data down into the Notch block. The Notch block was moving the camera inside its 3D world that allowed us to "raster" the output which was projected on top of the proxy tabletop. It was projected at a deformed angle that was only visible from the camera. So this was a very early proof of concept. The tunnel going into the table became quite a strong dynamic later on. Our proof of concept – the first images I made – was true to form to the final output, and that's partly because the environment I worked in was disguise's and Notch's shared 3D language; from the get-go to the end product the process was basically the exact same.*

Lewis is a firm believer that the main part of content is storyboarding, and double and triple checking the direction those storyboards and subsequent images involved might take. Lewis took Tawbox's 2D mood board images and basically sculpted a 3D event-based language.

> *We took Amber and Bronski's original first choreographed camera move from Cinema 4D, and we put that in the background. We worked out that we needed a surface for the top of the piano and we needed walls for the interior of the piano. In fact, we settled on only using two sides of the piano because as we rotated round we were only going to ever see two–three sides at a time. We then sequenced all this on a timeline so*

we had content from the music video, the storyboards, mood boards – all in one place. So if you were working in Notch, disguise Designer, or Cinema 4D, this was what you looked at to tell you what frame you were on, what part of the track you had to do – it was the backbone of our pipeline."

See Figure 5.10 for details.

> *We built various systems to do things – we built a pipeline from Cinema 4D into Notch, we built a pipeline from Notch into disguise, to Mo-Sys, also from Cinema 4D to fake Mo-Sys so we could work with tracking data while not in the studio. At the same time, there were separate pipelines working which was After Effects working solely on 2D content.*

All this had to happen in tandem to output a final image; that image had to be seen by camera and then sent to broadcast.

One early problem for Lewis and his team was that they didn't know what tracking system was going to be used initially, before Pod made the call for Mo-Sys to be used. However, they needed to get on with the job and start animating.

> *I took the data from Bronski and Amber's Previz in Cinema, sent it via MAX MSP using OSC (Open Sound Control) and then into disguise. I then transformed all that data into a language that could be sent over a network. That allowed me to fake a tracking source based directly from the camera Previs. We then passed said data into Notch block, which then in turn rendered the content into the perspective map via disguise.*

Essential to Lewis in working with real-time content production and 3D is clean up and preparation.

> *On that core level, you've got to know what you're doing – you can't just throw any geometry into real-time renderers and expect them to work. We were given a Brit Award [model] that looked like it had been scanned and had millions of polygons so we couldn't load that, it was going to chug the whole system up straight away and eat resources we didn't want to give to it. We had to go through everything and unwrap vertices so that it would work the way we wanted it to work.*

The team had a 2D workflow that was putting UV textures into the geometry, so a lot of the driving force of the narrative actually came from that 2D texture.

Another key aspect for the Stout Studio content team regarding working in real time is that it was important to avoid having too much laid down on a timeline.

> *We did everything externally and brought it in. We spent a lot of time in external software from Notch, mostly Cinema 4D, working on rigged character based stuff. We did a lot of keyframing. A lot! The hands going up to meet the prison bar was a fully rigged hand – we had to animate every finger, every joint . . .*
>
> *We were making changes on those key frames up to the last minute before the performance – the whole concept had to be designed with flexibility so that Bronski and Amber could make those last-minute adjustments.*

Figure 5.10 Technical Development on the Process working at Mo-Sys Studios. Source: Image Credit: Lewis Kyle White.

The End Goal (Not Modified to Be Successful!) and the Result

Despite the vast technical complexities that the concept of the performance posed, the end result was remarkably close to the original idea put forward by Tawbox. The O2 arena gave Dave a standing ovation, and the politically charged, ground-breaking performance piece of motion graphics content created headlines all around the world. See Figure 5.11 for this technology at work for the event and check out the full performance video at the companion website.

Dave's contribution at the awards ceremony was called by the musician Example "the most important performance in the history of the BRITs."

"I knew what was coming after two months of planning and rehearsing with a team of 16 but it was still so emotional on the night – especially seeing how much it connected with the young kids at the front and how much it meant to them. People were stunned. They couldn't believe what they saw," says Amber Rimell of Tawbox.

"It was an incredibly exciting and rewarding experience which ultimately inspired me to build my own XR studio in London for future projects of a similarly innovative nature," says Pod Bluman.

By Fiona Jennison for Bluman Associates https://blumanassociates.com/

At Bluman Associates, we are often challenged with making the impossible possible. Using technology in new ways to achieve different results, we design bespoke solutions, solving complicated technical problems with industry know-how. Our success is driven by our ability to listen. We're crazy about the visual environment and work closely with our clients to understand their audience and objectives.

Project name: Dave at The BRITs
Project location: London
Year: 2020
Project team:
Tawbox – Bronski and Amber Rimell: Creative Directors and Piano Design
Stout Studio: 3D Creation and Design
Shop/Tawbox: 2D Creation and Design
Bluman Associates: Technical Production
Pod Bluman: Technical Production Manager
Lewis Kyle White: 3D Lead Animator and Notch Designer
Catherine Woodhouse: 3D animator
Jan Urbanowski: 3D animator
Christian Dickens: disguise operator – pre-production
Vincent Steenhoek: disguise operator – show
Kyle Reseigh: Notch and disguise assistant
Stephen Ennis: Mo Sys
Ben Tilbrook: Mo Sys
Martin Parsley: Mo Sys
Jon Clarke: Steadicam Operator
Warren Buckingham: Focus Puller
Rupert Dean: Projectionist
Trevor Williams: Tour Manager (Dave)
Joel Stanley: Production Manager (Dave)
Client: Tawbox

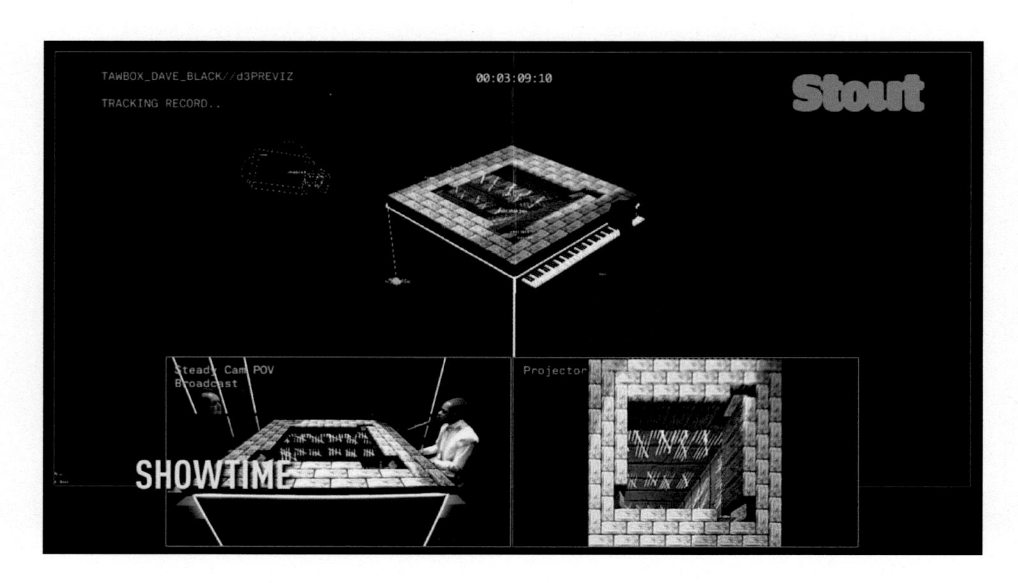

Figure 5.11 All systems working together for the performance. Source: Image Credit: Lewis Kyle White.

CHAPTER 6

Real-Time Content Generation Applications

We will review in more detail the many ways content creation teams can generate video in real time. Previous chapters have explored what we mean by real-time video content and the historical background leading up to our current toolset. We have also explored the ways teams use real-time content in a variety of disciplines. Here in Chapter 6, we take a deeper look at the different applications available for generating video imagery.

Custom Code

As outlined in earlier chapters, creators can write custom software to produce video content. Developers have a variety of video generation domain specific coding environments to choose from. Since developers can use virtually any modern computer programming language to create moving images, this is not a complete list, but a sample of three frameworks in current use within the creative video industry. Visit Chapter 3 for a reference to Blair Neal's article that provides a detailed look at creative technology resources.

Processing

> "Processing is a flexible software sketchbook and a language for learning how to code within the context of the visual arts. Since 2001, Processing has promoted software literacy within the visual arts and visual literacy within technology. There are tens of thousands of students, artists, designers, researchers, and hobbyists who use Processing for learning and prototyping."
>
> From the processing.org website

Processing provides a fantastic way to get started creating visuals with code for people both familiar and new to coding. You can download the language and work with tutorials online or go over to p5js.org and get started with a JavaScript version of Processing that works in your web browser.

> Go to rtv-book.com/chapter6 for active links to this example project.

In your web browser, go to this link: editor.p5js.org

```
Replace the code with:
function setup() {
createCanvas(400, 400);
}
function draw() {
if (mouseIsPressed) {
fill(0);
} else {
fill(255);
}
ellipse(mouseX, mouseY, 80, 80);
}
```

The result (referenced in Figure 6.1) is an interactive web-based content generation tool that works in real time. Move your mouse through the preview window to see circles defined by the ellipse statement: mouseX, mouseY for location, and 80, 80 for the ellipse size, making it a circle. Press the mouse as you move to fill the circle with the color black. Review the listed code again to learn how the mouse state defines the color.

These websites have many examples and tutorials to learn from. If you have more experience with coding or making devices with raspberry pi, processing.org has downloads for you to build projects. The companion website has some artist examples for inspiration and links to CCFest, the Creative Coding Festival, for more inspiration.

DOI: 10.4324/9781003206491-7

Figure 6.1 Processing code to make black and white circles. Source: Image Credit: Processing.org.

Next, let's review a programming language family called C that includes a language called C++. There are two language libraries built from C++ often used for creative coding projects, openFrameworks and Cinder. openFrameworks is described on the organization website as "an open source C++ toolkit designed to assist the creative process by providing a simple and intuitive framework for experimentation." While Cinder calls itself, "a C++ library for programming with aesthetic intent." Find links to both libraries on this chapter's section in the companion website.

If you are new to coding, creative coding is a great opportunity to learn computer programming with creative goals in mind. Many of the examples available online will share code and show in plain language how you can achieve the same results. Like Processing, openFrameworks and Cinder both possess a large community of users posting tutorials and example projects. Chapter 4 features a case study that uses openFrameworks.

Of course, using libraries or frameworks isn't a requirement for making generative art. JavaScript, Python, R, and virtually any generalized programming language can make real-time content. Readers proficient in programming should start with the language they know, and challenge themselves with a creative goal. Those with an advanced computer science background might want to dive into the latest developments in machine learning and procedural generation to create visuals.

Most importantly, what do you want to make? If the technology and problem solving interests you most, try partnering with someone more creatively inclined to come up with a project. Code won't make the art for you. It is just a tool, like a paintbrush. Sometimes a new paintbrush can inspire creative thinking or expand the creative goal. Creating art with computer programming comes from an artistic search for a medium that inspires the artist. Alternatively, artistic motivation might inspire you to learn to code. This exploration has no wrong answer as long as you make something.

Computer programming may not be the medium that inspires you, but an interest in video art may introduce you to a variety of software applications purpose-built for creative people who don't code. Many are designed for real-time content creation. Let's look at a few examples.

Effects Engines

Many software applications contain generative effects tools that create motion graphic image content in real time. Often referred to as effects engines or FX engines, they typically include a working environment with action descriptors and shape parameters to create particle emitters that look like sparks, smoke, waves, and more. Again, rather than a complete list, we will review a few to see what's possible.

Apple's Motion sells for $50. Motion provides an intuitive and simple interface that serves as a foundation for understanding real-time parameters for motion graphics and image generation. Similar PC applications, like Natron, provide a comparable experience, but for this discussion, we will focus on Motion.

Motion's chief advantage is a set of generators and behaviors for creating graphics and playing them back in real time while editing. This enables users to adjust parameters like speed, angle, replication, decay, and color while the project file responds to the alterations. In the process of making things, beginners can gain a solid grounding in real-time capabilities and terminology.

Visit rtv-book.com/chapter6 for download links and tutorials

While a host of applications can generate sophisticated motion graphics (After Effects, Nuke, Maya, Houdini, etc.), very few can play results in the editor without first running a pre-rendering process. While not designed for real-time output, as a working environment, Motion is resource efficient enough to show updates to most images in real time. Motion is a great primer before advancing to other, more complex, applications.

Media Server Effects Engines

As a video file plays, creators can augment elements of the video file to generate real-time effects. Alternatively, real-time effects can be wholly generated by playback tools. Media servers do a variety of video tasks from the simple to the complex. Servers can play video files in a predetermined order, map those video files to 3D objects, and generate content from files of executable code. A media server effects engine combines the organization of file playback with the capabilities of a content generation.

On video file playback, time is predetermined. A 30-second loop should always play in 30 seconds. Effects engines alter the files on playback while maintaining consistent playback speed. These effects can include altering color, limiting the number of colors used, adding noise or sparkles, or even reducing the image to blocks of color. The same effects tools can combine two video files. Early media servers depended on effects engines and stock content loops to provide new looks based on pre-made media.

As media servers advanced, so did the power of the effects engines. Custom made creative video content replaced stock content and content generators within the media server augmented the looks. A great example of this is the "Bugs" module on the disguise platform. The user defines bug density, turns, and speed. The result is real-time content using an on-server generator. Figure 6.2 shows artwork using the disguise Bugs module from Chema Menendez (middle), XR Studios (bottom), and Simon Anaya (sharing the love, top right).

The media server contains code defining a "bug" (not the software kind) and how one can alter its behavior. Users access controls available through the media server GUI to alter those behavior parameters so the desired result is created. Exposed parameter variables are a necessary tool in project files that allows users to easily alter content while in rehearsal.

You can find video effects engines and content generators in many popular media servers on the market today including Hippo, Pixera, Mbox, VYV, and SMODE. Please go to the companion website for links to learn more about these media servers from their manufacturers.

Let's look at some of the software suites used to create real-time controls and node based programming.

Node Based

Computer programming involves writing code to create actions, events, and behaviors desired by the programmer. What if those behaviors become so useful and repeated, that we used a common block representing each behavior instead? That is a rudimentary description of node based programming.

Instead of writing code, each node represents a self-contained snippet of code. Users arrange, graph, modify, and sequence these nodes to create their desired result. Here's a generic example using Vuo from an example that comes with the software (see Figure 6.3).

Vuo can be downloaded from vuo.org or at this chapter's web page and is an excellent way to get started with node based software.

Node based programming is a more attractive solution than code or effects engines to some content developers. Node based systems provide a great deal of flexibility for creating unique visuals while relying on pre-written sets of rules in the node blocks. Most node software systems also allow for customization by coding your own nodes. These systems enable users to build looks quickly and test variations with more ease than coding from scratch.

For image generation, there are many applications that use node programming that are regularly used in entertainment production. MaxMsp and vvvv are comparable to Vuo for content creation, while TouchDesigner and Notch are platforms that can handle more complex real-time content creation demands.

Figure 6.2 BUGS! Generative image content using a module on the disguise platform. Source: Image Credits: Chema Menendez, XR Studios and Simon Anaya and Damian Byrne.

You can find links for these applications and tutorials at the companion website.

This list is not exhaustive and new tools appear all the time. Wikipedia lists dozens of visual programming languages, but not all are ideal for real-time content creation. If you find an application not listed here that works for you and the way you like to create, that's great. Software applications have personality, some you'll enjoy working with and some you won't. Make sure the tool you choose has an active user community to share ideas and help solve challenges and you'll be in great shape. If a community doesn't exist, build one.

Game Engines

Another real-time content creation application is a game engine. Similar to node based visual programming applications, game engines function through the use of pre-built code modules to produce behaviors within a game environment. Most game environments need a 3D world, a player POV, lighting, and some consistent, if not plausible, laws of physics. Game creators can alter this pre-built coded architecture to meet the needs of their game's world. These tools have grown so powerful and reliable at rendering complex graphics in real time that game engines have found their way into entertainment content creation. The two regularly used game engines are Unreal and Unity.

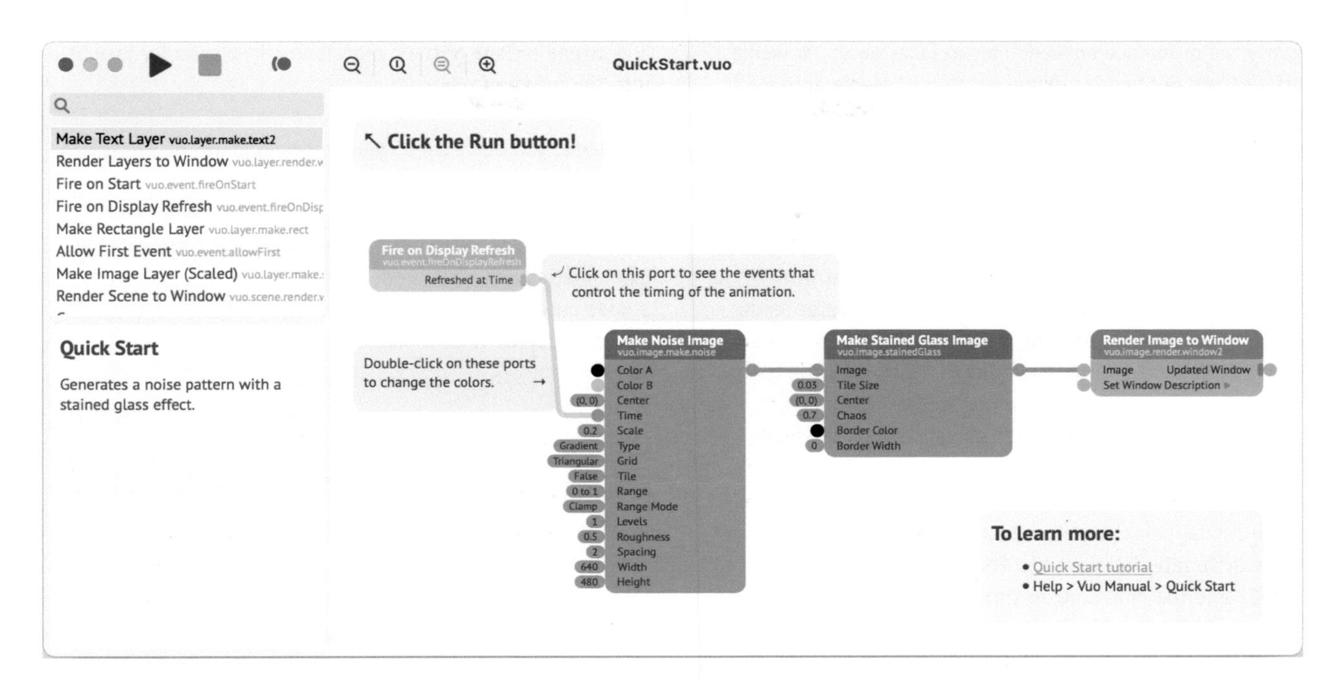

Figure 6.3 Example of Vuo node graph. Source: Image Credit: Vuo.org.

The Unreal Engine by Epic Games debuted in the late 1990s with the release of the first-person shooter game, Unreal. In the 20-plus years of development since, developers have used the game engine to produce not only cross-platform games but also real-time pre-visualizations of 3D environments. Unreal has emerged as a powerful real-time content generation tool.

Unity launched in the mid-2000s for MacOS. Now cross platform, teams use the development environment to produce game content for desktop and mobile. Unity is also used for pre-visualization and anywhere real-time 3D content manipulation is needed.

Anyone can download Unreal or Unity and learn these tools for free. I recommend some knowledge of 3D modeling, but both environments support 2D content creation. If you are new to 3D modeling, I recommend starting with Blender to learn the basics. This is a free software tool with a large user base and great learning resources.

Beyond Computer Graphics

All the real-time content creation tools discussed so far rely on computer graphics processing. The code, nodes, and engines described to create an image or video use physical knowledge of image processing. Using these tools, we describe color, shape, lighting, motion, and point of view to produce an image. Even when randomization or interactivity help generate a result, a user sets the rules for how the computer responds.

What happens if we allow the computer to guess at some rules? Can we create the desired content when all the states of color, size, lighting, and position are estimated or learned by the computer software? More importantly, can we improve the use of the available processing power to create more complex visuals in real time?

Machine Learning (ML) is a computer science discipline that trains algorithms to complete tasks based on examples of the correct result of that task. Successful ML algorithms are trained on large enough data sets to "learn" the correct or best outcome to a problem given brand new inputs within the same domain using the same data structures. For example, I can create an algorithm to identify a dog in pictures. To train the algorithm, I collect a large number of highly varied images, some with dogs, some without. Humans (via services like Amazon's Mechanical Turk) tag all the images that contain dogs. Next, I train the algorithm on a subset of my image collection to identify dog pictures. Then we test the algorithm with another subset of our images. We continue this training and testing process until the algorithm achieves our desired competency for identifying pictures of a dog. Similar algorithms identify the difference between human faces on social media platforms and photo apps. When your social media account prompts you to tag yourself in a photo, a machine learning algorithm found you there because it had enough photos to know how to recognize you in a photo it had never seen in a training set.

So if an algorithm can identify a dog, couldn't that work in the other direction? Can I say "dog" and have a computer render a dog for me? Neural rendering is an emerging image content creation technology. While a lengthy discussion on the topics of Machine Learning, Deep Learning, Neural Networks, and Artificial Intelligence is beyond the scope of this textbook, these advances will have huge implications for the world of content creation, and specifically real-time content creation. Though I must point out these technologies are advancing at an incredible rate. Between the first draft of this chapter in fall 2021 and the final draft in spring 2022, OpenAI launched DALL-E 2 which can not only draw a dog for me, but can draw a dog with a pet tiger in the style of a 1960s rock poster just based on text input. Check out the video at the companion website.

Unreal and Unity both rely on procedural content generation to create game environments. Procedural content generation uses ML algorithms to create imagery based on a range of provided assets and rules for how to combine them. Defining landscapes of fields, mountains, cities or other planets is time-consuming and resource heavy to store. Why not teach the game the landscape look and have the application generate more of that landscape for you as needed?

Another place we see ML at work is directly on the GPUs powering these engines. From the GPU developer Nvidia on their DLSS technology: "Deep Learning Super Sampling is groundbreaking AI rendering technology that increases graphics performance using dedicated Tensor Core AI processors on GeForce RTX™ GPUs. DLSS taps into the power of a deep learning neural network to boost frame rates and generate beautiful, sharp images for your games." And what's good for games is good for entertainment real-time content creation.

> Check out some discussion of machine learning in content creation at rtv-book.com/chapter 6

As Machine Learning tools advance, real-time content creation will incorporate these technologies. Changes happen fast and it can seem overwhelming to keep up with the latest graphics and content creation technology. I recommend following SIGGRAPH and attending one of their conferences in person for insight into the future of our work.

Of course, you don't need computational power or degrees in AI rocket science to reach your creative goal. Bleeding edge technology can produce mind blowing results, however old-fashioned analog solutions still generate beautiful real-time manipulations of video. Let's look at a case study combining a node based content creation tool with a clever code-free manipulation of video signals.

Case Study – Electronic Countermeasures – The Foo Fighters 26th Anniversary Tour

The Power of Physicality and Presence in Real-Time Effects

The stages had been dark for a little over a year when we met with production designer Dan Hadley to discuss the direction of the video design for the Foo Fighters' summer 2021 tour (see Figure 6.4). Masked, vaccinated, 6 feet apart at a picnic table beside the Los Angeles River, freshly sanitized hands passing a sketchbook back and forth, as we envisioned what would be possible for the first major tour to return to the road since the Covid-19 pandemic shuttered the entire industry.

We oscillated between being clear-eyed about the challenges and being starry-eyed about recovering the spaces and experiences we loved. Another wave could shut everything down again. The people, gear, and infrastructure needed to support a tour had been lying fallow for 18 months. The pre-pandemic machinery of live events didn't shut down gracefully and would take a lot of time and effort to spin up again, but what an honor and joy to be the team that got to pick up the pieces and bring audiences and performers back together again.

More than ever, we wanted to feel the thing that we had taken for granted prior to the pandemic. We wanted the collective energy of a large audience all focused on a moment in time, becoming part of the music, the band onstage becoming part of each person for that moment. That connected presence we had suddenly lost was ours to create again. After 18 months of virtual, distanced, and telepresent performances, the sound waves from cheering fans were going to rattle each other's eardrums alongside the music. Light reflecting off Dave Grohl and the band was going to strike the retinas of a crowd celebrating the return of Rock and Roll; live and in person would be the core concept and guiding principle for our design.

Figure 6.4 The Foo Fighters on tour. Photo Credit: Emery Caleb Martin.

We wanted to focus on the light and sound that are already present in the arena, to bring the audience closer. To give them access to details that rarely make it onto an IMAG screen; closeups of the textures of a live show, like pedal boards and strobe tuners, and the architectural intricacy of drum kit hardware. To make them part of the design itself. So, instead of bringing in screens content, we'd capture everything live and build a video instrument to manipulate all the input into a cohesive look for each song that would be responsive to the individuality of each performance.

Live and real-time have always been key components of any video for the Foo Fighters. They don't play to a click track. We don't get to script out the show with timecode for an infinitely repeatable performance because the band loves to improvise. They rearrange songs on the fly, sometimes weaving together several familiar melodies into one extended bridge. From the perspective of the video operator, it can be beautiful chaos. Complexity has to balance with flexibility and performability because we never know exactly what is coming next.

Continuing the metaphor of the video instrument, if we wanted to play along with the band, we needed to think about the visual elements as sonic components: keys on a piano or drums in a kit, or perhaps more accurately, pipes in a cathedral-sized organ. The player selects each element to create the melody, harmony, rhythm, and dynamics of a piece of music. The challenge of this metaphor for video and effects has always been one of effective parameterization and control; narrowing the nearly infinite possibilities of images captured by a camera and pixels on a screen into an effective set of keys that one operator can reasonably manipulate. Meaning that the design work for this tour was going to be twofold, not just the visual output to the screen, but the interface and user experience design to make this instrument usable by the programmer and operator.

First the basic hardware: The Foo Fighters already tour with their camera package of two pedestal cameras, a steadicam, a rail-mounted Panasonic PTZ on the drum riser, and a PTZ over the drum kit. We added four PTZs and two AJA Rovos to that package, with control over those six cameras routed to the video programmer's station. Video playback is a pair of disguise vx4s running Notch. The control surface is a GrandMA3 console.

Next, the software: A Notch master-block we wrote to enable live camera switching inside a look and a standardized set of parameters for easily controlling the individual video manipulations built into the composition for each look. Within the master block, each song had a layer that served as its particular look. The base functionality was a clean pass through that allowed for switching between the program feed cut by the Foo Fighters' director Josh Adams and the 11 individual camera feeds. Layers for other songs included projecting multiple camera feeds onto a ring of 3D curved surfaces or using hand-drawn textures to matte the program feed of the band over the live feed of the audience.

The possibilities offered by this approach were endless and luckily we had time for a lot of play and discovery while developing the looks that would ultimately appear in the show. Even though we did this aspect of the show in a real-time engine, development of looks followed a very traditional path, including sketching on paper, and mocking up ideas in a traditional compositing or 3D environment before translating them into Notch. Stacks of hand drawn frames waiting for capture to use as masks and mattes filled every spare flat surface. Flowcharts and state diagrams to help visualize the various control elements each look would need were pinned to the walls. The studio looked like a tornado had picked up a software company and an animation house, spun them together and dropped it in our laps. We usually enjoy our jobs, but the build process for this tour was something truly delightful.

A fully software approach, however, wasn't quite enough to meet the brief of celebrating a return to live shows. Most times, the effects offered by real-time software lack a certain personality and sense of physicality and naturalism that we prize in our work. The Foos needed something a little more real, so they commissioned us to build four Physical Effects Boxes (PFX) to perform real-time video manipulation in physical space.

We developed the initial PFX concept and R&D on the Beck Night Running tour, responding to the artist's desire to perform inside a mirrored room, but without having to tour with a mirrored room. Realtime ray-tracing to create live reflections in software wasn't readily available, and still isn't nearly as satisfying or accurate as real mirrors. The PFX solution was born from an offhand comment of "this would be so much easier with a shrink ray . . ." and even though we can't quite miniaturize a performer and put them in a tiny, tourable, mirror room, we can send a video feed into a tiny, tourable, mirror room. See Figure 6.5 for an example of the internal build of one of the PFX devices.

Figure 6.5 One of Electronic Countermeasures' PFX devices. Photo Credit: Emery Caleb Martin.

Essentially, the PFX boxes begin life as standard 4U rackmount computer chassis with power, SDI video in and out. Open the case, however there's no computer inside, it's a tiny maze of mirrors, cameras, screens, and other optical trickery, which reflect, refract, slice and dice, and transform an incoming video signal using the power of actual old-fashioned physics.

We designed four different effects for the 2021 tour:

Acute (Figure 6.6): a sharp narrow triangular, bottomless corridor made from four angled mirrors.

Wavirror (Figure 6.7): a rectangular mylar portal with haptic motors to translate sound into vibration of the mylar and pixel addressable RGB tape to color the mylar to match the stage lighting.

Tryptych (Figure 6.8): an open faced mirror box split into thirds using a mixture of two-way and normal mirrors.

Shatter (Figure 6.9): a broken mirror pieced together by hand and captured by two cameras with two different focal points,

Figure 6.6 The Foo Fighters concert stage using the Acute PFX Box. Photo Credit: Emery Caleb Martin.

Figure 6.7 The Foo Fighters concert stage using the Wavirror PFX Box. Photo Credit: Emery Caleb Martin.

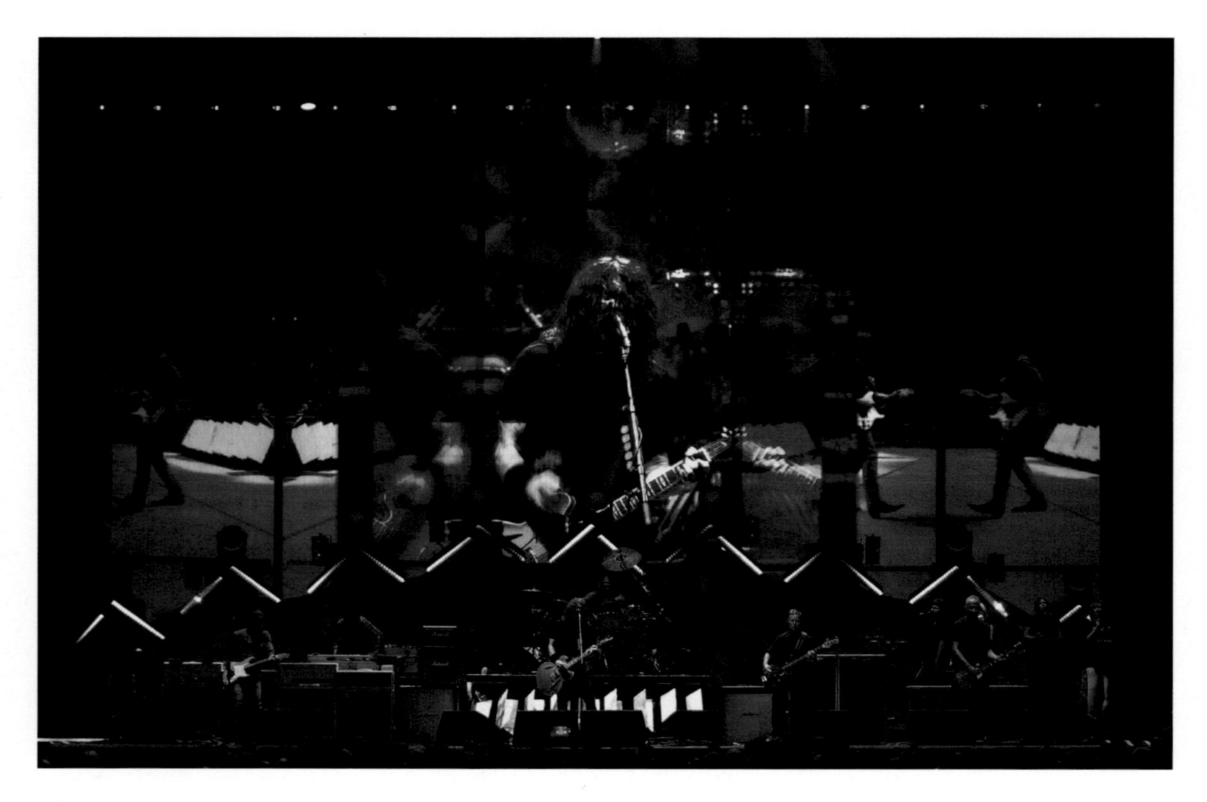

Figure 6.8 The Foo Fighters concert stage using the Tryptych PFX Box. Photo Credit: Emery Caleb Martin.

Figure 6.9 The Foo Fighters concert stage using the Shatter PFX Box. Photo Credit: Emery Caleb Martin.

one with sharp focus on the video reflection and the other on the cracks of the mirror.

The PFX boxes, as designed, add a level of naturalism and serendipity to manipulating the image that we don't see digitally, even with as far as we've come with rendering technology; the nuances of the imperfections of dust from the road, the misplaced fingerprint, and the serendipitous camera framing that causes a reflection off one surface to bounce and catch an imperfection in the material of another to flare at its edge makes the magic even if you don't know why.

Taking a song like "The Best of You," putting an audience in a position to be staring into an LED screen that's become distorted, broken, and disfigured by an actual broken mirror, changes the tone of an IMAG screen and especially so in the middle of this pandemic period where everything and everyone is a bit broken. The emotional response to the PFX boxes is subtle but palpable. Figure 6.10 shows the internals of The Tryptych PFX Box.

The Pipe Organ that is the Foo Fighters' video instrument is continually evolving to better serve the creative and functional needs of the show. We have sketches for new looks, ideas for more elaborate physical effects, streamlining the hardware and software for lower latency, better paradigms for control that allow for more responsive programming and performance.

Listening to the Foo Fighters play live, especially when they're riffing on the songs and artists that have inspired them, weaving motifs from Rock and Roll history into their own songs as they shape their own Rock and Roll story offers us an unparalleled inspiration for designing the future of real-time video.

By Kerstin Hovland and Emery Martin, Electronic Countermeasures
In memory of Taylor Hawkins and Andy Pollard

Project Credits
Electronic Countermeasures LLC
www.ecminteractive.com/

We are a studio of post-disciplinary artists creating immersive experiences and environments of all shapes, sizes, and

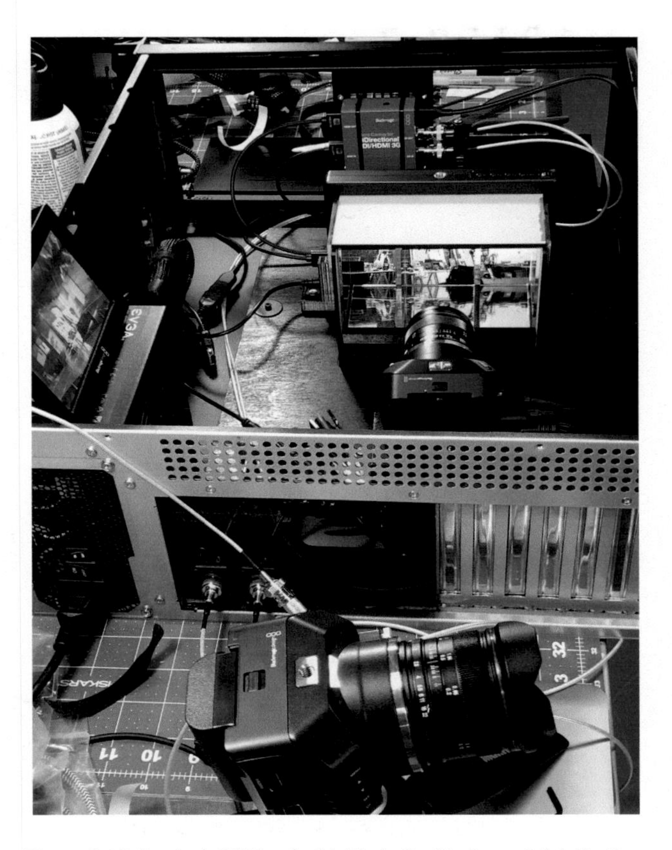

Figure 6.10 Tryptych PFX box build. Photo Credit: Emery Caleb Martin.

mediums. We come from a diverse background of creative, academic and technological study including art, music, theater, art history, critical theory, animation, film, computer engineering, virtual reality, human computer interaction, design strategy, and digital fabrication. From projects tiny to enormous, hand-built to technologically daring, Electronic Countermeasures welcomes the chance to amaze, delight, and create unexpected and unforgettable experiences.

Project name: The Foo Fighters 26th Anniversary Tour
Project location: Music Performance Tour
Year: 2021
Project team:
Hayden Katz – Video Programmer
Josh Adams – Camera Director
Kerstin Hovland – Video Art and Design
Emery Martin – Video Art and Design
Client: The Foo Fighters

CHAPTER 7

A Collision of Technologies

The first chapters of this textbook review the growth of computer graphics, broadcast graphics, and real-time content creation tools. Only a handful of these technologies were developed with the entertainment industry in mind. Instead, we borrow and repurpose tools, informing the industries in which they originate and pushing their growth into new creative disciplines beyond our own. Technologies collide in the entertainment industry to create new and unexpected experiences.

The collaborative nature of entertainment production means people with a wide range of skills work together and learn from each other. As a result, practitioners bridge tools and ways of using those tools to meet design challenges, while expanding their personal knowledge base. We adapt tools from many disciplines around us to advance the technology behind the entertainment spectacles. Every production offers us an opportunity to interact with new people with parallel skills and different experiences, and to learn from them.

What Are All the Tools We Combine?

Entertainment technology is an established discipline. There is likely someone at your university teaching theater technology or technical direction. However, creative video production for live entertainment is a young field. No common protocol or standardized workflow exists for the use of video content in entertainment production outside the theatrical discipline of projection design.

> I wrote about creative video production and technology for live entertainment in "Screens Producing and Media Operations: Advanced Practice for Media Server and Video Content Preparation." Visit rtv-book.com/chapter7 for more information.

Real-time content use in entertainment is a rapidly developing practice. We experiment with how to execute as often as we reuse plans that have worked on prior projects. This textbook can't give you a perfect outline of the future, but it can provide a knowledge base of what has come before, share an understanding of the current technology, and create insight to where those technologies might take us. From there, the best skill is the ability to combine available knowledge and reliably use it in new and interesting ways.

Real-time content creation does not exist in a vacuum on a production site. Many other video technologies and creative video practices are still in regular use. These tools now integrate with real-time applications and should be considered partner technologies. Let's look at the capabilities and origins of various entertainment production tools used alongside real-time content creation.

Media Control and Management

Creative video production teams have used media servers for well over a decade to fill screens and video scenery. In this market, media servers are known for their playback flexibility and ease of managing and mixing files together for output to non-standard screen sizes. These systems typically have effects engines, playback manipulation tools, and image warping functionality. We have discussed media servers in prior chapters with a list of current manufacturers available in the companion web documentation for Chapter 6.

In the early 2000s, media servers could play back a few video files down one or two 1024 x 768 signal paths, and maintain the desired frame rate. These systems typically delivered signals to projectors or low-resolution LED devices. Today's more complex descendants of those original media servers output 16 or more HD feeds, synchronized to broadcast genlock and SMPTE timecode, all while managing image mapping and scenery tracking.

DOI: 10.4324/9781003206491-8

Media servers arose as video proliferated in scenic design. Early use of projection suffered from low light output and complex image alignment. Stage lighting levels can overtake the lumen output of projectors in a live performance environment, requiring time-consuming level balancing. Combine this task with projector alignment, often an overnight endeavor with multiple people at work, producers and production teams alike considered projection a luxury for scenic design. Projectors have improved dramatically in brightness, contrast, and resolution in the last 10 years. Many media servers now automate projector alignment through the use of structured light or IR beacons, making projector use far simpler and more cost effective. As a result, we see projection on more and more productions.

LED screens have also seen major advancements that popularized the use of media servers. Low resolution LED tubes, spheres, and screen panels filled with stock content debuted in sets in the early 2000s. In 2010, we considered an 8mm screen a very good resolution. Today, 2mm is a very good resolution. Building a set entirely out of LED tiles became feasible by the mid-2010s and now LED screens offer such high quality that they serve as backgrounds in film production.

More scenic surfaces are dedicated to video, with some now made entirely of LED screens. The expansion of video scenery has not always come with an increase in production time or other resources. To be effective within such constraints, media operations teams depended on the speed and flexibility offered by media servers. While these teams refined their processes, media server developers responded with tools to meet the growing production demands. These tools became critical to managing file augmentation in rehearsal and quickly making creative changes to rendered content. Such content alterations might be temporary, later fixed and re-rendered in the video content, but increasingly, these alterations became the final product.

Effects engines have limits to how much they can alter a rendered video file. While some media servers have generative content engines built in, many media server manufacturers support third party real-time content creation tools, rather than building their own. By hosting third party content creation software, media servers can generate the content defined by these executable files.

Generating real-time content along with the existing media server tools on the platform creates a powerful combination. Traditional media server functionality, like cue structure, file management, and traditional media playback, remain critical to live entertainment and have valuable uses in working environments dependent on real-time content production.

Projection Mapping and Scenery Tracking

Simply put, projection mapping is the craft of precisely positioning an image pixel on a given surface. The mapped surface can be anything: a 16x9 projection screen, an LED wall, a bank building downtown, or an inflatable dinosaur. While properly aligning a projector to an inflatable object may sound challenging, sometimes locating content correctly on a flat LED screen can be no less difficult. Both depend on software tools to define the relationship between image and output.

As mentioned earlier in this chapter, media servers can do much of the work of properly relating pixels to their correct output location on a screen for both projection and LED workflows. Projection on dimensional surfaces tends to be the most demanding, requiring accurate 3D models of the projection surface, as well as accurate position location and lens information for the projectors. Content is virtually mapped to a 3D model within the media server. The media server uses the simulated projector as a virtual camera pointed at the object. This "camera" captures the content mapping from the projector point of view, generating a video signal of the result (see Figure 7.1). Once this video signal is delivered, the projector outputs in the physical world the corrected imagery that was simulated in the virtual world of the media server.

Scenery tracking occurs when a video surface moves and the LED or projection output must respond to the change of location. A 16x9 screen can be flown vertically or tracked horizontally across a stage to reveal new content. That inflatable dinosaur can rotate, requiring image correction for any content projected onto the dinosaur surface. In both cases, the media server needs real-time information that defines the location and orientation of the screen surface relative to the location of the projectors. A number of different technologies are available that can track scenic position and provide data to a media server. This data enables accurate video content placement onto a moving surface.

For complex surfaces, position, orientation, and speed are combined with a 3D model of the projection surface hosted on the media server. The media server interprets scenic position information, matching the object model to the physical state of the scenery. Imagery is corrected in real time as the simulated projectors/virtual cameras within the media server passively "view" the result. If the tracking information is accurately interpreted, generating the corrected video content is the same

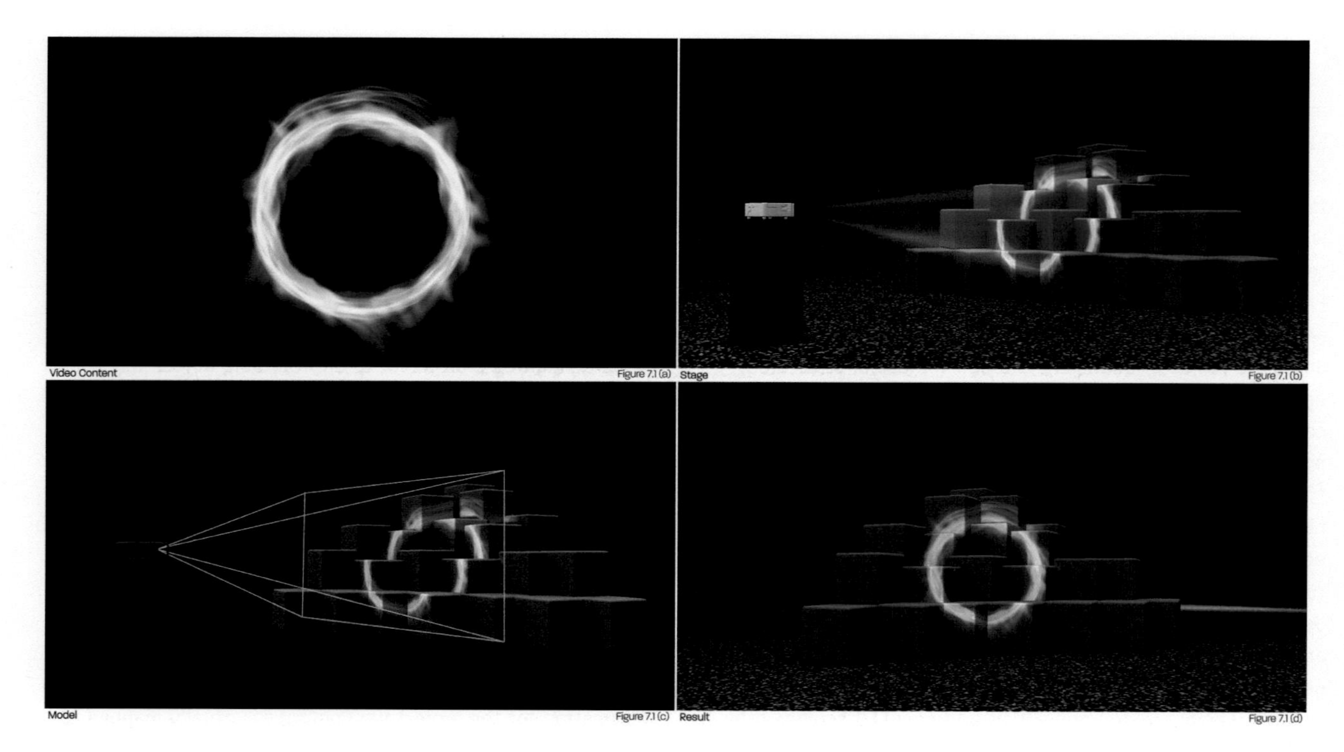

Figure 7.1 This series shows content projected onto a model so that the result is cleanly mapped onto a complex 3D surface. Source: Image Credit: Ian Macintosh.

process for moving scenery as it is for static scenery projection mapping. Media servers rely on infrared (IR) beacons or scenic automation data to locate the tracked projection surface. We'll see this same tracking technology used to track people and cameras.

Environment Modeling

I cannot overstate how critical good 3D models are for the success of real-time content generation. Accurate models of the scenery and venue are necessary to correctly model the real world into the virtual world for content generation. We rely on the precise locations of those models relative to the exact location of cameras, projectors, and any IR beacons receiving data to correctly output perspective accurate content.

The entertainment design process frequently relies on 3D models, however, the model as designed might not exactly match the set as built. This can impair the content correction quality and tracking capabilities of a media server. Models often need correction onsite or must be recreated altogether.

Various platforms offer different approaches to model correction. Some provide ways a user can correct the model within the media server using onboard object editing tools. Instead of returning to the originating 3D modeling tool, editing the model within the media server can show the result live while the corrections are being made. Other platforms assist with model correction employing IR beacons embedded in the scenery or tagging important vertices with the server interface mouse.

If a new model must be created onsite, there are a few approaches. Some media servers can generate a new model using structured light calibration tools. Externally, photogrammetry can be used to create a model from still photographs of the object. For the most accurate result, LiDAR (Light Detection and Ranging) will create a point cloud mesh using laser pulses that can be converted into a model.

The more accurate the model, the more precise the final image result will be. How that model is achieved is a factor of the time and budget you have available. Projection mapping and scenic tracking are only some of the use cases of model data. As we will see, the 3D model of a stage environment is foundational to the success of current real-time content use.

Camera Tracking

Stage design has long depended on tricks of the eye to create depth. The point of view of the audience is controlled into a narrow seating angle, making it possible for perspective

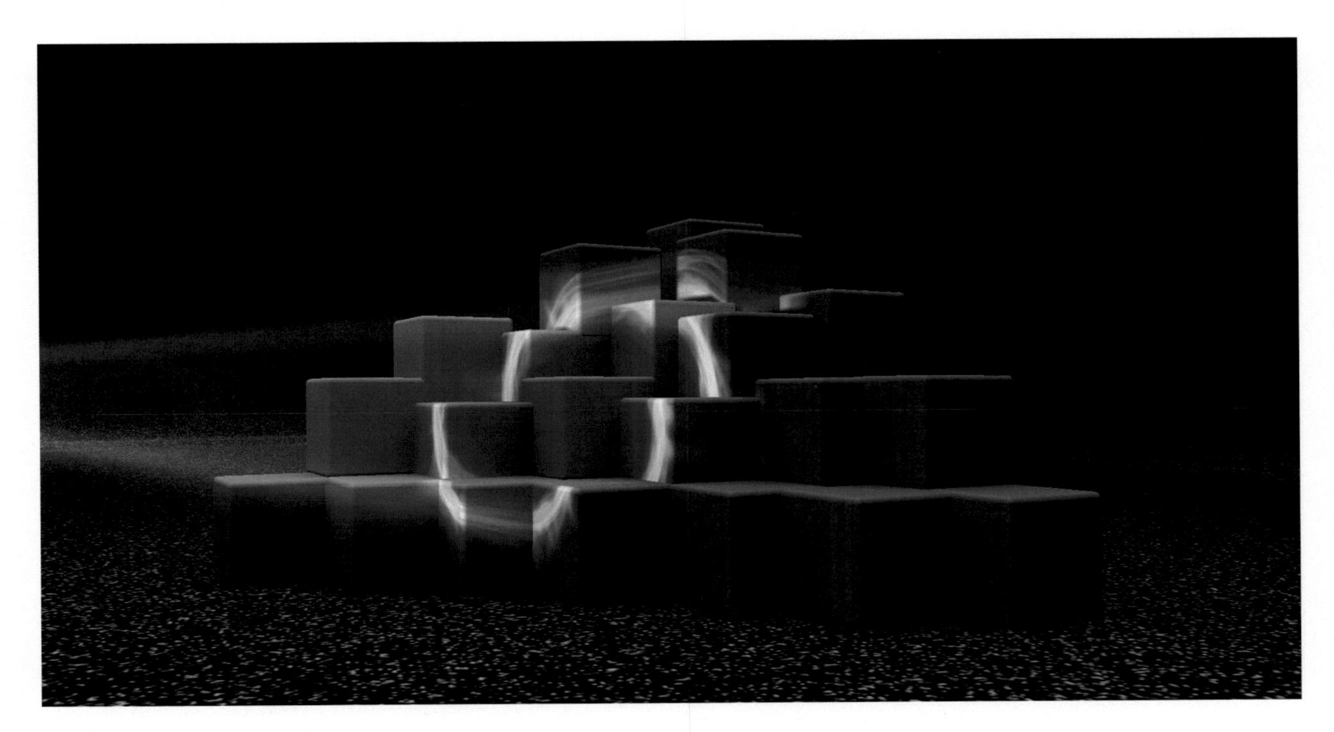

Figure 7.2 This image references a view of the stage from Figure 7.1, showing the scenic projection approximately 40 degrees off axis from the projector. Source: Image Credit: Ian Macintosh.

painted scenic drops that give dimension to a flat surface. The same tricks are employed with video content. Projection mapping creates depth with forced perspective visuals. When viewed off angle, the content will appear broken or skewed (see Figure 7.2).

In film and broadcast, the camera acts as the viewer, delivering the captured image to an audience on a TV, computer, or phone. Lens position, angle, and zoom of the camera directly affect the appearance of any content with forced perspective dimension. For the perspective to look right, imagery must be created to the point of view of the camera. With real-time content creation, if we know the camera location and status, we can adjust the content to match that camera perspective live.

As with scenic tracking, encoders inside a camera assembly can feed data to a media server. When camera mechanisms cannot not provide data, optical tracking tools can locate the camera in space, similar to the systems used to track people and scenery. This works well for both handheld camera and steadicam rigs.

Alternatively, we can pre-program camera movements using robotic control systems. Sports broadcasters use robotic systems to capture high speed racing footage or to place cameras over a field of play. In film, robotic camera assemblies follow pre-choreographed camera moves to facilitate complex shots that are repeated again and again with a physical camera for principal photography, and a virtual camera for VFX post processing.

Camera tracking and control technology has had decades to mature in other use applications. For entertainment, we combine that camera data with the pre-existing technologies we have reviewed here to generate content in an entirely new way. The most significant of these combinations is the use of the camera tracking data to update the status of a virtual camera in a 3D model.

As described in the section on scenery tracking, we noted that we "view" a moving object in a 3D stage simulation with a virtual camera to correct projected video content on the physical object. When camera tracking is employed for content creation, we match the virtual camera to a physical camera inside the content creation platform. The movement of the virtual camera captures a simulated 3D scene to create perspective accurate content for the tracked physical camera (see Figure 7.3). Same tools, new process.

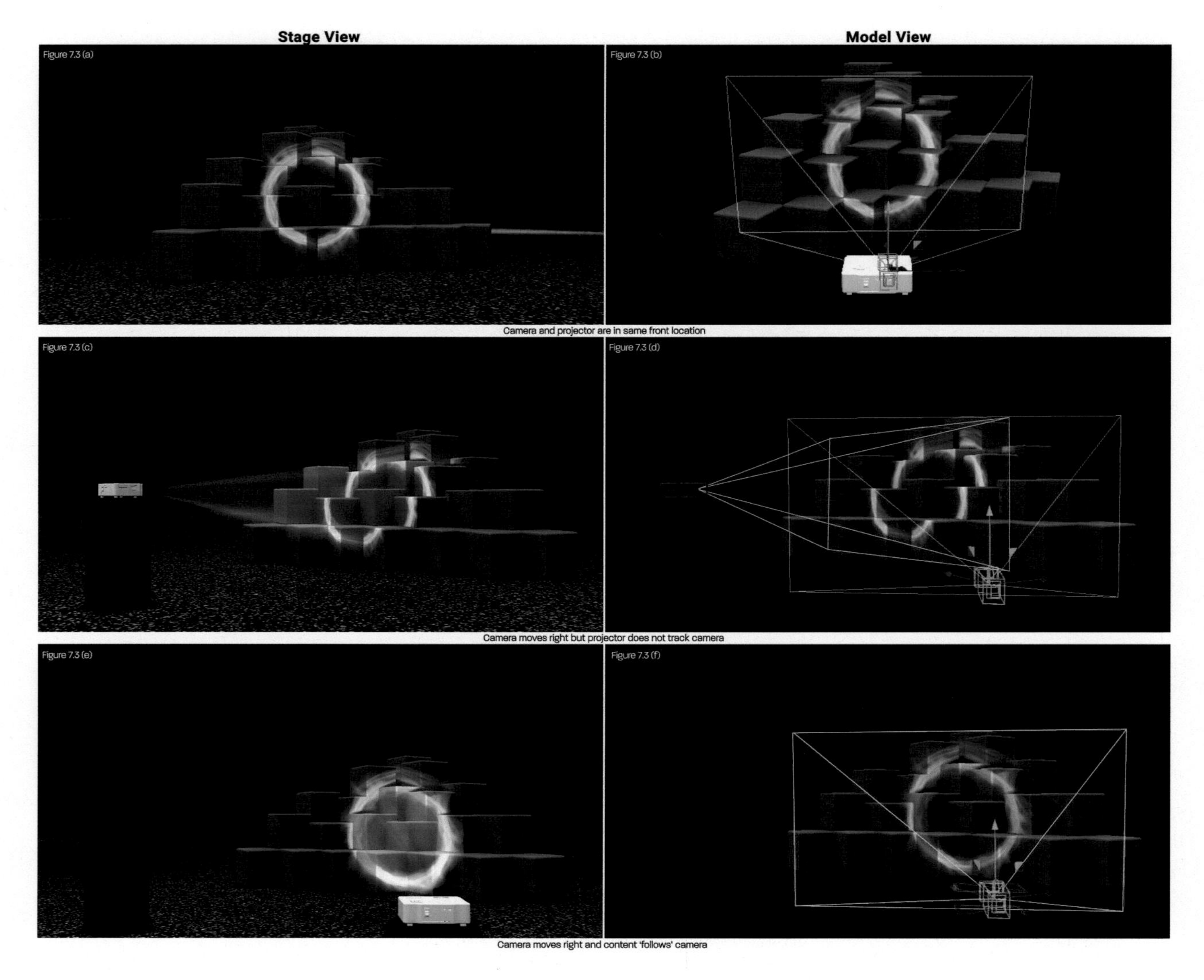

Figure 7.3 This image series describes a process in which a camera views the stage from center and then shifts to a view from house right. When we relocate the projector to the new camera position, the video content is then corrected for that stage view angle. Source: Image Credit: Ian Macintosh.

Content Generation

When we combine media servers, scenic and camera tracking, and accurate model information with real-time content creation, we get a whole new generative scenery workflow. In prior chapters we've discussed a few use cases and will continue to review these combined technologies in more detail. First let's look at the introduction of real-time content creation for live entertainment use.

Chapter 6 covers the different ways content can be created in real time. These tools were in limited use in live event production until the introduction of Notch in the mid-2010s. At that time, Notch was seen as a way to rapidly introduce interactive control to content creation and improve the rendering pipeline process. These speed efficiencies made interactive content viable for the short schedules experienced in live entertainment production. As user knowledge expanded, more exploration was made of what variables and data sources could be combined with this content toolset. This included the use of position tracking data and its use in generating content relative to the location of an object.

Let's create an example of a performer wearing a hat with an IR beacon. Notch receives location data from the IR trackers located around the stage to define the hat's position within the tracked space. As the performer crosses from Stage Left to Stage Right, a trail of sparkles follows them in a screen upstage, as if the hat has magical powers. To make this sparkle effect work, a set of Notch blocks generates sparkle imagery and locates this content in the output based on data from the IR tracking system and a model of the stage. Meanwhile, the media server hosting the Notch file manages and maps the output to screen.

Next we'll add camera telemetry to our example. With this information, we can match a virtual camera within Notch to a physical camera following our performer. As the physical camera moves along with the hat, content is generated that matches the perspective of the camera point of view and the hat location.

Now let's enhance the camera footage. We can take the camera feed and composite sparkles in front of the performer at the exact location of the hat. Again, the hat and camera are both tracked within our 3D model inside of Notch, which then generates foreground content in real time to match the camera perspective and composite the result. When we marry the foreground imagery, or front plate, to the camera footage, we are using Augmented Reality (AR will be reviewed in more detail in Chapter 9).

Visit rtv-book.com/chapter7 for explainer videos from textbook contributors, Bild, about Broadcast AR.

Let's go back to those sparkles that are visible in the screen upstage of the performer. What if we build a 3D model of the place where the performer is walking, like an old village road? There are trees and stone houses as well as the hat sparkles. As the physical camera on set follows the performer, the matched virtual camera captures a different perspective of the old village model and generated sparkles. This real-time generated scene is then delivered to the LED wall as a video signal to be visible to the physical camera at the correct point of view. When we deliver perspective generated content in real time, via a blackplate, to a background of a live action scene, we are using Background Replacement (we will review Background Replacement in more detail in Chapter 10).

As described in Chapter 5, Background Replacement is a dynamic digital scenery tool that makes use of all the technologies discussed to create a virtual backdrop to a performance. Productions began experimenting with this technology in the mid-2010s, but Background Replacement workflows are best known from the example of Disney's *The Mandalorian*, produced in late 2018.

Check out the clips posted at rtv-book.com/chapter7 for more insight into the Background Replacement technology used on *The Mandalorian*.

Let's look at the various steps taken to create *The Mandalorian* LED stage environment. First, the virtual scenic environment was designed in a real-time software tool, in this case, the game engine Unreal. A media server was used to host a 3D model of the stage as well as the content creation software. Next, the stage camera is precisely tracked and matched to a virtual camera in Unreal. The modeled imagery in the field of view of the virtual camera becomes the digital background we call a back plate. The back plate is mapped to screen by the media server to the exact perspective of the physical camera and delivers the needed video signals for the LED stage walls. This process is continuously delivered in milliseconds.

No one built any one of these tools for this express purpose, but brought together, they create something uniquely powerful for perspective generated backgrounds. As we have seen, we can create camera perspective generated content in the foreground of a camera shot and in the background screens in real time. The relationship of a virtual camera in a 3D model to a physical camera on a stage is the cornerstone of all virtual production practices.

Therefore we can define virtual production as a real-time VFX defined by the relationship between a physical camera and a virtual camera. The virtual camera exactly mimics the attributes and behavior of a physical camera in a simulated 3D environment, generating perspective sensitive content separated into foreground plates for AR workflows and background plates for Background Replacement workflows.

In an LED Volume, think of the camera in the VP process as both camera and projector. The physical camera informs the location and lens of a simulated camera in a virtual environment. This environment is typically the digital scenery composition for the background of the shot. What the camera captures in the virtual space is then "projected" onto a model of the screen configuration. This information is then converted into the signal paths for the physical screens in the studio. The physical camera in the studio captures perspective corrected content. Use the earlier example of a set of projection mapped

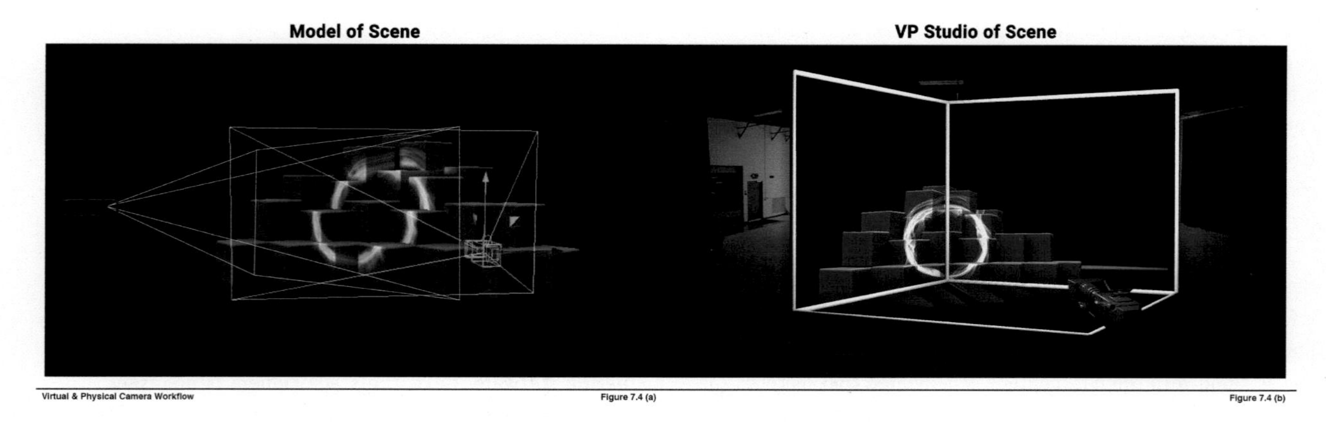

Figure 7.4 Take the stack of boxes from Figure 7.3 (d), align the projector to the camera and output to an LED stage. Source: Image Credit: Ian Macintosh.

boxes and imagine they are now the background of our VP scene (Figure 7.4).

Front Plate and Back Plate

Let's review our definitions of AR and Background Replacement from Chapter 5. AR is the compositing of camera sensitive, computer generated images in the foreground of a live camera shot. Background Replacement is the real-time generation of camera sensitive, computer generated images in the background of a live camera shot that is either key composited with the camera foreground or live captured in the camera background. XR, or eXtended Reality, occurs when both AR and Background Replacement are combined. AR, Background Replacement, and XR are all real-time content dependent subsets of Virtual Production.

We need the following for AR, Background Replacement and XR to function:

- Duplication of the camera telemetry from the real world into a virtual computer generated world in real time
- Accurate representation of the physical world of the shooting space in a 3D model
- Generation of content assets in real time
- Real-time image processing or compositing depending on the plate usage
- Video signal delivery

The Augmented Reality image generation workflow produces a front plate of camera perspective sensitive content that is overlaid onto the camera signal. The imagery sits in front of the live captured footage. Content is produced with transparency information defining the AR imagery and the area that passes through to the camera captured content.

The Background Replacement image generation workflow produces a back plate of camera perspective sensitive content that is for use behind the live action footage. Depending on the stage style, the back plate is delivered to an LED stage for live capture by the camera, or composited to the camera signal with a chroma key.

When AR and Background Replacement are combined for XR, the live captured footage sits in between the front plate and back plate (see Figure 7.5). Some systems can generate both plates simultaneously, but it's considered best practice to separate plate production onto synchronized real-time content host servers.

The term Virtual Production covers a host of technologies depending on the needed plates for your project. How the plates are combined with live footage will depend on the VP stage type.

Virtual Production Stage Types

LED Stages/Volumes

Virtual Production stages employed by film-style productions like the *Mandalorian* use an LED screen ceiling and surround primarily for live Background Replacement. In a Volume, the floor is often a practical environment with physical props and other scenery, though an LED floor can be used. The LED walls can be designed to fully enclose the stage to provide reflections from all directions.

The background, lighting and reflections captured in the shot are referred to as In Camera VFX, or ICVFX. While the content is live generated based on camera position, these stages are not common for live broadcasts. Volumes are favored by film productions that allow for a post-production process. Any real-time compositing is done to generate a previsualization of the final result.

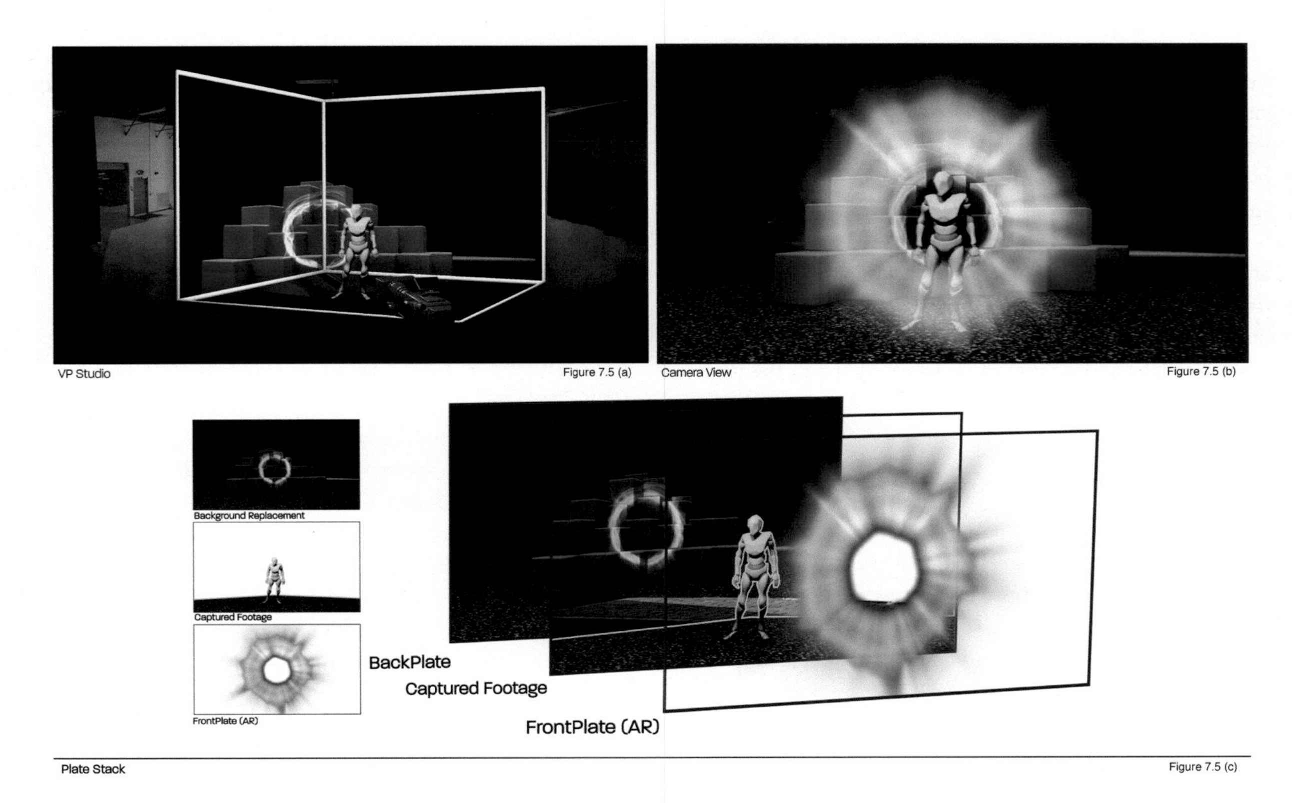

Figure 7.5 Foreground and Background plate structure. Source: Image Credit: Ian Macintosh.

LED Stages/XR Stages

XR is shorthand for eXtended Reality or miXed Reality which is the combination of AR and Background Replacement. XR Stages are typically more compact than LED Volumes and include an LED floor. Ceilings are left open for practical lighting, though they can include some panels to provide natural reflections. Camera shots expose the edges of the LED wall to be managed live or in post.

The smaller LED format can be extended virtually using the same content generation tool used for the LED wall content. Referred to as Scenic Extension, this complex process creates an AR layer of background content.

XR stages dominate in markets shooting live to broadcast or that need to capture as much as possible in principle shooting with limited post production processing. Check out the companion web chapter for clips of XR stages in use.

Modern Green Screen

Virtual Production tools are used for modern green screen production. As previously discussed, productions have used green screens for decades. However, camera tracking has improved the capabilities of Background Replacement when using green screen.

Green screen remains a far more cost effective working environment when compared to LED Volumes and Digital Stages. Full virtual Background Replacement with chroma key also eliminates the issues that come from seaming together the LED walls with virtual scenic extensions.

The primary disadvantages of using green screen are the loss of natural looking reflections from the emissive walls and the ability of performers to engage naturally with their surroundings. Green or the chaosen key color can also show up in reflections requiring post production clean up.

Online Events

Online events are an entertainment technology that experienced accelerated development during the pandemic. While audiences could not gather in person, live events continued bringing their audiences together through online portals like Zoom or games like Fortnite.

The primary difference between an online event and a Virtual Production live broadcast resides with the audience. Online events take advantage of the interactivity of the web through gaming style avatars and interactivity. I expect we will still see entertainment online long after live performances get back to pre-pandemic levels.

As of this writing, in early 2022, live event production has started coming back, but many of these online audience Virtual Production technologies remain in use and will continue to develop. We will explore them all in greater detail in the next section of this textbook.

What Happens Next?

No one can absolutely predict how teams will use these tools in the future or across different sectors in the entertainment industry. Production companies and creators continue to install LED stages all around the world. Popular music artists perform within online games. We see a steady growth in demand for people with skills to support these endeavors.

Hybrid Audiences

Fully remote audiences can take part in virtual events as viewers navigating a game, or as avatars interacting via game style controls. Not everyone wants to attend an event in person, but many want to experience as much as they can through virtual services. Every day we can see more events furthering virtual participation through hybrid audiences. Part of the audience will attend in person, while another part will attend through a remote viewing option.

Remote Performance

If part of the audience can view remotely, can performers deliver remote performances? LED stages and synchronized cameras can put two performers next to each other in a broadcast, when they are in studios hundreds of miles apart.

AR for Live Events

AR depends on cameras to work, so how do we use AR tools in a live performance? Since most audiences will have a device in their pocket with a camera and screen available, these can augment a performance from the audience perspective.

Creative video and real-time content creation have benefited significantly from these varied technologies coming together and creating a new production experience. While many of the examples we've discussed require expensive and complex gear, you likely have access to tools to experiment with some of this technology right now. Check out this TouchDesigner + AR case study from Rich & Miyu.

Case Study – Rich & Miyu – AR Livestreamer

A TouchDesigner Based Software Solution for AR Applications

AR Livestreamer is a software layer built on top of Derivative's TouchDesigner that allows for placing real-time 3D objects in the real world. We developed the system during the pandemic to allow our clients to make their livestream shows and events more engaging without expensive hardware.

The system itself is a collection of tools and creative parameters collected together into one user interface. See Figure 7.6 for reference.

The system allows for cameras and Azure Kinects to be connected and calibrated to a scene and then different objects can be triggered as cues to appear in world space. It can then stream out this feed via RTMP to YouTube or Twitch, etc.

Our two biggest challenges were:

1. Calibrating the cameras to match the virtual scene to the real world
2. Embedding objects with shadows to make them look like they're really embedded in the scene

Calibrating the cameras required building a system using a checkerboard alongside the OpenCV libraries' camera calibration functions to enable us to calculate the lens intrinsics and extrinsics.

First, we calibrate the camera's intrinsics using the checkerboard and its corners (see Figure 7.7). This helps us to calculate the lens radial distortion in a camera and then undistort the lens to give us a flatter image.

Learn more about OpenCV at rtv-book.com/chapter7

With that approach, we can then use the points in the checkerboard to calibrate the camera's intrinsics (the projection matrix of the lens) and finally its transform position in the real

Figure 7.6 The AR Livestreamer Show Control interface. Source: Image Credit: Richard Burns.

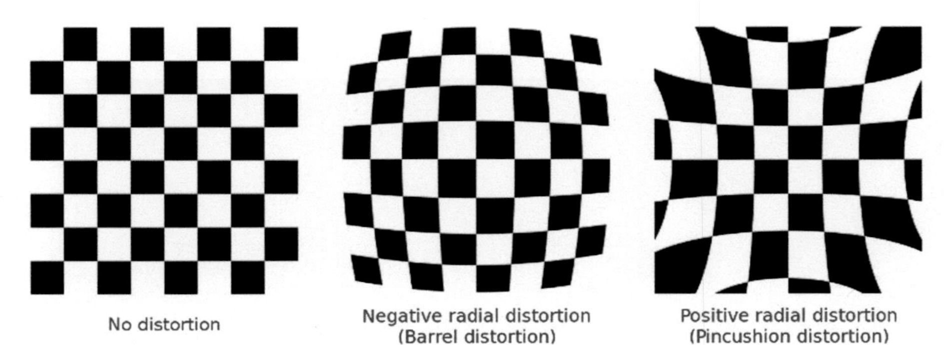

Figure 7.7 Checkerboard Calibration Grids. Source: Image Credit: Open CV.

world. See Figure 7.8 for our first successful camera calibration using this method.

For our second challenge of making objects look real, we needed to give the object an environment that matched the real world. We found that if we took the camera and applied it to virtual materials using environment mapping we could get real reflections that reasonably match the environment of the room (see Figure 7.9).

We then matched this environment map with a single source light that we could move around our 3D scan space to create a simple shadowing effect (see Figure 7.10).

To cast a shadow, however, we require a plane to cast the shadow on to. We allowed for positioning of this plane in the system and made an automatic plane that would sit on the floor.

Rendering and compositing all these elements together gave a reasonably realistic effect.

To finalize the look, we add a slight grade across the entire image and around 0.5% additional noise, which helps embed the virtual object in the scene and makes it feel a little more filmic. Figure 7.11 shows the resulting composite of camera and AR image.

The checkerboard calibration approach worked well in controlled environments, but for larger spaces and moving cameras we needed another approach. We decided for larger environments we could use a Kinect to build a rough point cloud of the 3D space, making it easier to align distant cameras.

Figure 7.8 Rich & Miyu testing camera calibration. Source: Image Credit: Miyu Burns.

Our technique with the Kinect was to place the Kinect in the middle of the space and then take 3D snapshots as we rotate the device, gradually building a full 3D scan of the room (see Figure 7.12). This could be done with a LiDAR scanner for larger areas, but that would be more expensive and most locations we were planning to use the system were live-house venues or clubs.

Another benefit of having a Kinect is that we could allow the entire scene to become virtual and then use a joystick to move our camera around that scene. The system allows for a mix of pre-scanned static point clouds and live Kinect based point clouds. The joystick could move on paths and then bring itself back to the original camera position before blending back into the real world. Figure 7.13 shows the virtual space on the left and the real space on the right in the camera AB mixer.

This system is in ongoing development and there are constantly new technologies being released that can assist us with the system. NVidia released a system for automatically green screening people without a green screen which makes masking objects behind performers much easier provided they are

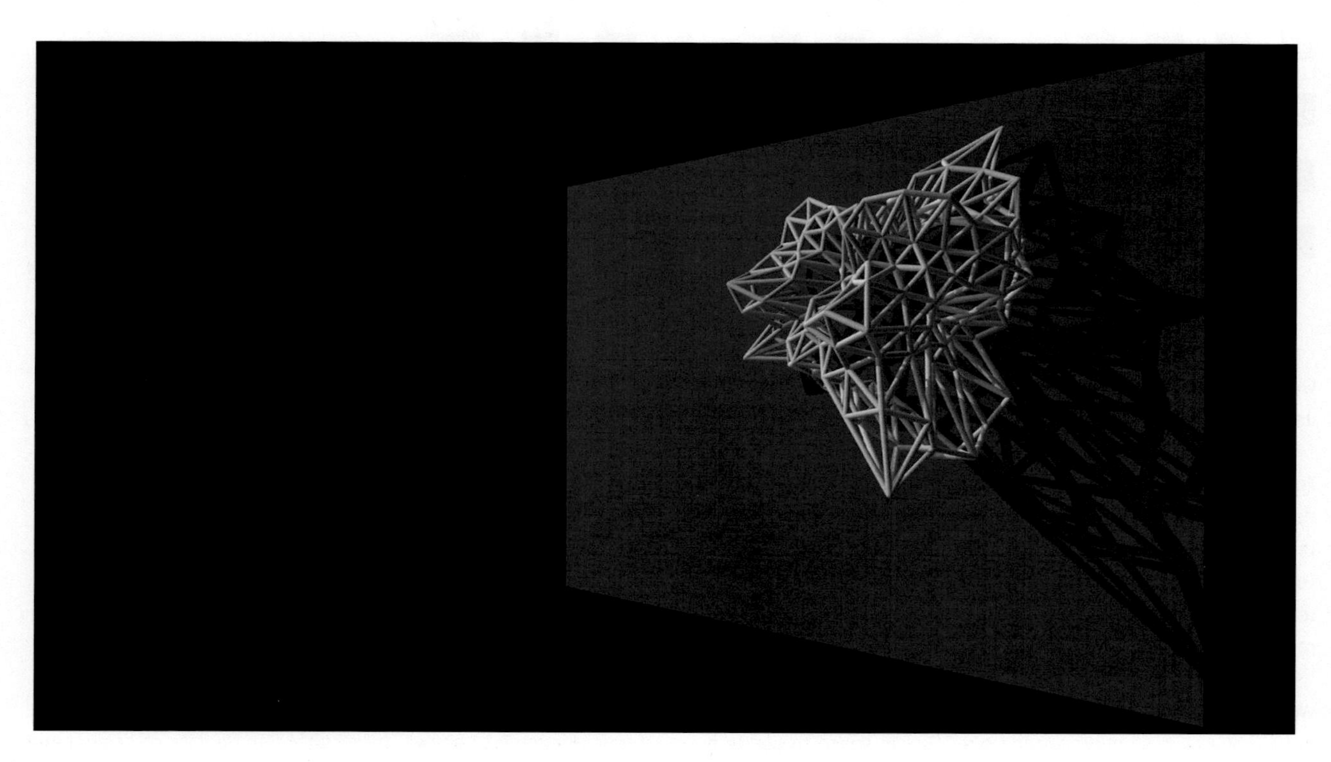

Figure 7.9 The original beauty pass render, with shadow plane. Source: Image Credit: Richard Burns.

Figure 7.10 The shadow intensity map used in the render. Source: Image Credit: Richard Burns.

Figure 7.11 The final composite with all elements of the scene. Source: Image Credit: Richard Burns.

Figure 7.12 A scan of the Rich & Miyu office with a couple of basic virtual objects in the scene. Source: Image Credit: Richard Burns.

Figure 7.13 AB mix of cameras. Source: Image Credit: Richard Burns.

face-on to the camera, removing our need to use a Kinect for this purpose and allowing us to go back to a camera only approach for simpler applications.

Project Credits
Rich & Miyu
https://richandmiyu.com/

Rich & Miyu consists of Richard and Miyu Burns. We provide creative and technical services in both English and Japanese and work in the field of interactive design, art and research. We are based in Tokyo but work on projects internationally.

Project name: AR Livestreamer
Project location: Tokyo
Year: 2021
Project team:
Richard Burns (Design and Development)
Miyu Burns (Graphic Design)

SECTION 2

Real-Time Content Entertainment Applications

CHAPTER 8
Content Production

What is the motivation to use real-time content creation? Many software applications enable the creation of motion graphics and narrative video content. Every tool requires time to learn and gaining mastery of a tool often requires years. Is it worth learning new tools? Why switch from your current content production workflow?

This chapter will look at the various factors that make real-time content production the best choice for a particular production. Before we go any further, I do not intend this chapter as an argument for (or against) using real-time content. Generative content tools have many advantages when used in the right working environment, but not every environment is right for these tools.

To begin, we must clarify the difference between real-time content manipulation and real-time content generation. Both are real-time content tools. Content manipulation implies the existence of a video signal or pre-rendered video file that is enhanced in real time. Content generation assumes code and assets combine to create a video feed in real time. We will see examples of both kinds of content production.

Like any other creative video software application, using real-time tools requires planning and preparation time. We must consider the design requirements and production type when selecting content production tools for any project. Sometimes traditional content creation tools might be the right choice. Understand the goal and use the best tool. Here are some factors to consider when deciding between traditional versus real-time solutions.

The No-Render Advantage

Real-time content creation tools eliminate the rendering process. This time-consuming step (or set of steps) of traditional content creation software can add hours to the time it takes for a team to see the results of a small change.

For example, a 3 minute motion graphic file built in a tool like Adobe After Effects can take 10 minutes to render or 90 minutes to render depending on factors like file complexity, computer capability, and codec compression. Changing any of these factors could reduce render time, but that's not always feasible.

Once rendered, we must deliver a file to the team responsible for playback. This may entail file upload to a cloud service and then download on the receiving side. Depending on the file size, the exchange could take minutes or hours. In fact, moving files to a hard drive and sending a physical device to an onsite team via overnight delivery may prove faster in some situations.

Once delivered, the onsite media operations team must check and distribute the files to the playback hardware. File transfer speed depends on network speed, and everything from network switches to cable quality affects bandwidth. An improper network configuration or the use of low-quality cables can result in transfer speeds well below typical expectations.

Without discussing the content requirements, we can see that rendering time, file delivery, and distribution issues already make a case for pursuing a more time-efficient alternative. Real-time content files are often quite small, in the kilobyte range, instead of the multiple gigabytes that are typically required of a rendered video file. Existing as lines of code makes delivery and distribution of these files fast and easy. For purely motion graphics content generation that relies on off-site content teams and internet delivery, rendering time alone might be sufficient reason to support a move to real-time content tools.

Files heavily dependent on secondary assets like complex 3D models and video textures will not always benefit from distribution efficiencies. Assets like these equate to an

DOI: 10.4324/9781003206491-10

unavoidable increase in file size. With proper network planning and fast internet access, cloud management will be your best option.

What the cloud cannot provide for real-time content is asset playback. Most cloud distribution services for rendered video can preview the clip. Some platforms can previz rendered video content mapped to a model of the stage. However, preview of real-time video content files online is not possible for most file types. Currently a real-time content file will need to be rendered for creative review and approval. In time, I expect we will see "no-render" solutions for real-time asset review available via web browser.

Creative Flexibility

Another common use case for real-time content creation tools is creative flexibility. The clear advantage of real-time is the ability to make detailed adjustments to content and seeing the result immediately. Don't like the number of clouds? The placement of a tree? The direction the birds are flying? Turn a knob and instantly the birds are going the other direction. Ok, not quite like that, but it could be like that if we included a bird direction variable and assigned it to that knob.

When using traditional content tools, someone must detail and deliver change notes like these to the motion graphic artists and editors. After making the changes in the content software, results are re-rendered, re-delivered, re-distributed before the playback team finally plays the new file for creative review. This cycle repeats until the team realizes the creative vision or the project must go live (too often it is the latter).

With a real-time content tool, teams must make edits in the software, but distribution of the update is smaller and faster. Alternatively, we can plan for content adjustments in advance by incorporating control variables into the code. We can alter these variables as the content plays with the results visible immediately. The number and distribution of birds, or the position coordinates for the tree can change in real time if we construct the content design that way.

Knowing which variables to plan for can be a challenge. Attaching adjustable parameters to every element in a real-time project isn't practical. Variable use requires thoughtful planning as part of real-time content design. Typically, the content team returns to the source file for more complex edits or major redesigns. Real-time is powerful, but that power comes from good creative decision making.

I tell clients real-time content creation tools allow for faster iteration and more opportunities to refine the content, but the creative process leading up to production remains the same. Real-time content creation still requires a creative brief, team discussion, client approvals, and content production. Creative flexibility doesn't mean a team can show up on a project with real-time content generation tools expecting to create new content on the spot. As with any type of content production, we cannot escape the need for planning and prep time.

Live Signal Manipulation

Real-time content creation tools can be utilized to enhance live camera feeds. As discussed in prior chapters, we can generate content in response to camera position or we can manipulate the video signal itself. Real-time content creation combined with powerful computing to maintain framerate is the only way to achieve live alteration of a camera feed. A computer hosting the real-time content toolset receives and processes the camera feed. These tools alter the signal, frame by frame. For signal manipulation using an effects engine, the change covers the full frame, altering color or image. Real-time content software like Notch offer more detailed image manipulation control and generative content additions to the signal during processing. This can include graphic overlays, facial recognition and embellishment, as well as other full frame artistic image processing.

Live alteration of IMAG (Image MAGnification) is a common use of real-time content manipulation. In a live performance with a large audience, cameras and screens are used to make presenters more prominent. Dedicated IMAG screens located around or within the stage are used to project the camera feed, enlarging the events on stage for better audience visibility. IMAG feeds are often altered to match the performance creative.

Real-time camera feed augmentation also occurs for broadcast. Camera feeds can be composited with generated graphics or scenery, as discussed when using Broadcast AR. These overlays can be anything from a 10 yard line marker in a football game to a virtual space shuttle opening a portal through the venue ceiling and landing on stage. Most use cases for Broadcast AR will make use of camera telemetry to place graphic imagery into the frame (we will discuss this later in this chapter).

The live signal may be another video content file. During playback, further enhancements are applied in real time to manipulate the look of the content using media server effects engine processing or real-time content creation engines. This process is useful for enhancing existing content, or mapping rendered content to complex or moving surfaces.

Interactive Content Production

Another common real-time content creation use case is interactive content. When a performer or viewer can affect the behavior of the visual display, the system must render the change in real time as the trigger event occurs. For example, we may track a performer across the stage in order to determine the position of generated sparkles on the screen upstage. Or the hand movements of a viewer, tracked with a depth sensing camera, might trigger a response in a video art installation. This might sound complex, but we regularly see examples of real-time content interactivity outside performance entertainment events.

Think about a simple video game. Video displayed on the game screen renders in real time in response to player interactions and input. This is so common, I would argue that many no longer consider this complex, but as described in previous chapters, decades of work have gone into game development to achieve this sense of effortless interaction. In game content creation, event triggers are highly controlled. Whether using a handheld controller like a joystick, mouse and keyboard, or a gamepad, those inputs have nearly universal definitions for game play. One button makes a game character run and another changes the player point of view.

The trigger inputs for a performance or installation are far more varied. You could design a trigger based on a sound input or on the speed of arm movement that initiates a change in the content. Maybe you create a content generator dependent on weather data in a city selected by the viewer. The input trigger, how the trigger interprets the event, and the video content generated in response to the event must all be custom designed for the particular context of the performance or creative piece.

Let's make up an example using sound as an input variable for video content. We have a bit of software that generates puffy clouds going left to right across a blue sky. In our installation, audience members can sing into a microphone. A sound analyzer will measure the pitch and level of the sound. Low pitch generates fewer clouds and high pitches generate more, while low volume maintains a slow speed and high volume accelerates the cloud movement. We won't leave any instructions for users participating in the installation, just the microphone and video display. Can you imagine the various sounds a person might make as they figure out how our interactive installation works?

Instead of a microphone, what if we just told the viewer where to stand? We can use a depth camera to analyze that viewer's movement. Using skeleton tracking, we can assign variables to arms and legs. Arm movement left-to-right controls cloud number and standing on one foot changes cloud color. An IR tracker placed on the viewer's head will track location and speed. We'll assign this data to cloud speed. Each of these tracking points can provide data to the real-time content creation platform as design variables for content generation.

Interactive content depends on the variables. Those variables must be designed as part of the creative process. For non-interactive content production, we use variables to facilitate the design process. Eventually, those variables remain static when the design is complete. When used for interactive content, variables become integral to the final design. Variable values remain unknown until a viewer or performer interacts with the content trigger. The result of that interaction changes the variable value and the behavior of the real-time content.

Be sure to check out video resources for this chapter at rtv-book.com/chapter8

Environmental Responsiveness

Interactive content can respond to its environment. For example, these same tracking tools used to locate a person in space can be used to locate scenery on a stage. As a projection screen or scenic piece moves, content can remain consistently displayed on the surface using real-time content mapping. Mapping requires the use of a 3D model to locate the projection surface and the projectors in space, and use that information to correctly align the content.

For example, let's imagine a bust of a head that is painted with a projection material. We can create any video content we like using any tools we like to project onto this head. A media server is used to map the content onto a 3D model of the head, and output the resulting correctly warped content to the projectors.

Now let's imagine the head rotates. The projected content needs to remain in place on a rotating object. While the resulting look on the object is the same, we must manipulate the content in real time to change the projection correction on the head surface. For this example to succeed, the platform managing the content and sending signals to the projectors must know the rotation position and rotation speed of the sculpture at all times. With this data, the system can update and correct projected content in real time to maintain the desired visual result.

Interactive scenic and performer position tracking are common forms of real-time content manipulation and generation. Besides IR and depth sensing cameras, scenic automation encoders can also deliver real-time accurate position information to the servers managing content output. We can employ these same tools to accurately locate the position of a camera and the status of its lens.

Camera Perspective Generated Content

Currently, one of the most exciting uses of interactive content production is based on camera tracking. While the camera can be a source of real-time (live) content for image manipulation from the camera feed, the position of the camera is also a critical data source for environmentally responsive content generation.

As discussed in Chapter 7, camera position and lens data are necessary to generate perspective sensitive content in real time. As the camera moves and changes its field of view, real-time content tools will interpret this data and generate content to the perspective of the camera viewing angle. We can use this information for dynamically generated backgrounds or 3D foreground imagery.

These spatially sensitive content resources depend on real-time content tools. The results are so useful that this type of content production will be the focus of the next several chapters discussing various types of real-time content dependent Virtual Production.

Whatever the motivation for using real-time content creation tools, there are challenges to overcome when adapting to a new way of working. An unfortunate side effect of the flexibility that comes with a "no render" solution is that "real-time" can be mis-understood to mean "right now." Production team members might not be aware of the complexities of their use.

For real-time content to meet expectations, crafting the code or software for the desired image result will be a time intensive pre-production task, similar to any content creation. However, the result is a highly flexible and interactive file that can dynamically respond to the needs of the production. In the following case study, we'll learn more about one team's process in adapting to a real-time content creation workflow.

Case Study – Fray Studio – Real-Time Update

How many hours have we all spent waiting for things to render? Watching frames slowly count upwards; it is like watching digital paint dry. In 2018 we were very frustrated with this practice and decided to try something different for *Back to the Future the Musical* (see Figure 8.1 for an example of visuals from the production). We decided to break up with After Effects and embrace a new real-time workflow.

At FRAY, we had all been using After Effects since we started working with video. It has always been there. It is the Swiss army knife of the video designer. We never considered there to be an alternative and had not sought one. We had our set of plugins, they did what they did, and all the work started to look a little too much the same. We were firmly entrenched in our workflows.

We were too used to telling a director or fellow designer that it would be tomorrow before we could have that great new idea rendered. We were very used to not being able to preview our work in real time despite After Effect's best efforts. We added Think Box's Deadline into our workflow: lashing together a network of computers into render farms to speed things up, which helped a little but it was fraught with its own problems and without a competent deadline user on the team it could become useless.

This was our professional world; forever waiting too long for renders while trying to make exciting things happen on stage in real time. The piece of software at the core of our workflow was not up to the job. If it had been a media server, it would have been out the door, and a new one found. But we never asked the same question of After Effects. It was the status quo. And everyone at FRAY was very bored with the status quo.

We had been increasingly using Notch for creating real-time content for media servers such as disguise but had not used it a great deal for making stand-alone video content. Our growing disillusionment with After Effects coincided with us speaking at a conference where Matt Swoboda of Notch was also presenting. His message was simple, Notch is a tool designed for live-video and After Effects is not, so why use it? Initially, this sounded like heresy, but the more we thought about the idea, the more

Figure 8.1 Still frame of video content for *Back to the Future the Musical*. Source: Image Credit: FRAY Studio, 2020.

it made total sense. There had to be something better out there.

The thought of breaking up with After Effects, like ending any long-term relationship, was a scary one, but we knew it was time to move on. We had a large project in the shape of *Back to the Future the Musical* on the horizon which we had already decided would have some real-time elements made in Notch so why not go the whole way? We set ourselves the rule that we would do our best to start and finish everything in Notch. We would only resort to After Effect or Cinema4D renders if we hit a total roadblock. Cinema 4D however was allowed to be used for all 3D modeling.

Overhauling our tried and tested studio system workflow was a challenge, but a worthy one that threw up many questions: how do we structure projects, how do you do something in Notch you do automatically in AE, how do you make your work accessible to others, is it rendered or real-time and how do you decide? See Figure 8.2 for an example of the Notch interface we used.

First and foremost, the rendered or real-time block question needed answering as the approach you take to making the content is very different. We decided on a few basic principles to guide us:

1. If the video content needed to endlessly generate something like a sky, it would be real-time.
2. Variables or interactive elements that would change continuously or at random would be real-time.
3. If the moment was locked to time code or did not respond to random variables, it would be rendered.

Building a content for real-time playback in the media server has a very different set of considerations to that if you are rendering it. There is a delicate balance between the level of detail in the look you are building and the performance of the media server. We spend a lot of time looking at the performance tab. Juggling between one significant smart node or several smaller, simpler nodes to find the correct balance of performance while maintaining a vibrant and exciting look. Furthermore, the real-time content might be mixed with rendered content while the media server is processing automation data, dmx tables, and any of the myriad of things they routinely do in running the show which means less power to deliver your blocks. This is slowly changing with the advent of the disguise VX/RX systems but those did not exist at the time and still are expensive and out of reach of many Westend/Broadway shows.

Once you have your blocks balanced with the system performance, the number of blocks in our disguise project

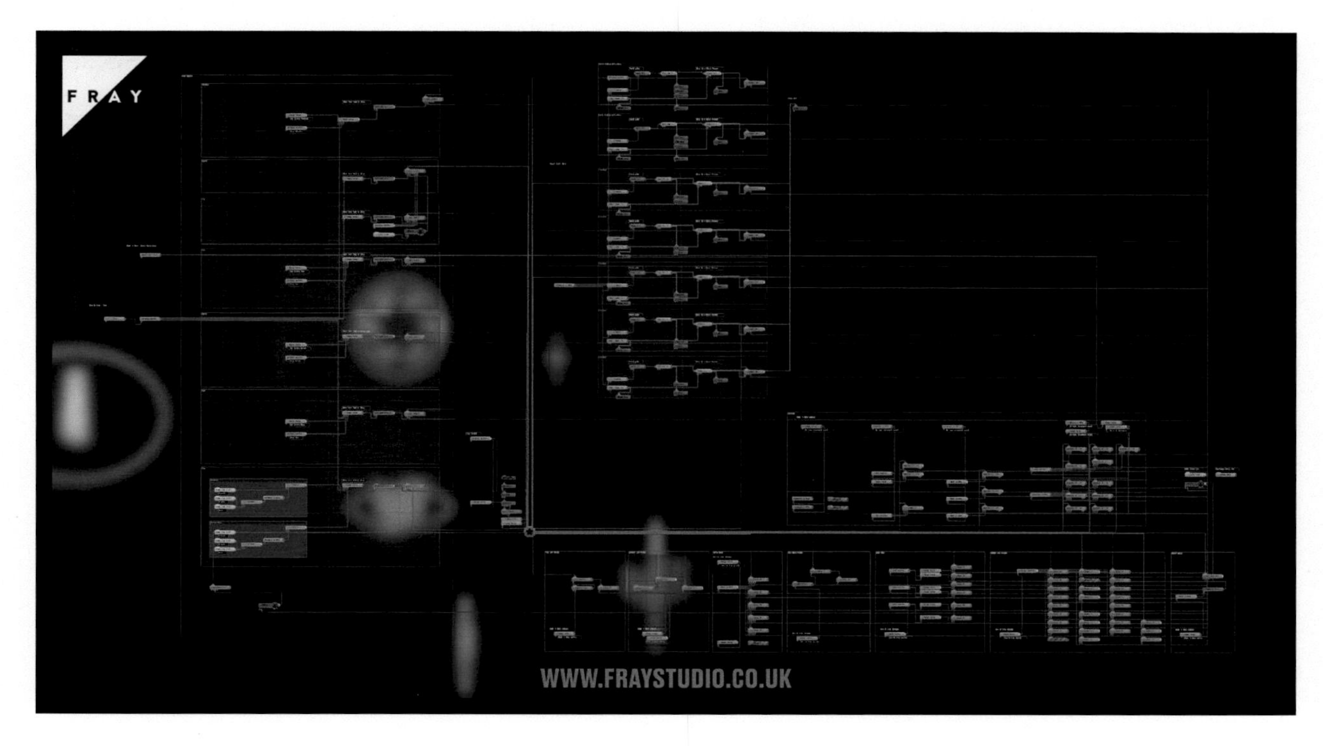

Figure 8.2 Example of Notch node graph for *Back to the Future the Musical*. Source: Image Credit: FRAY Studio, 2020.

also needed careful consideration. More blocks mean longer load times for your project and complex blocks add to this time. Furthermore, it is just a computer processing the block, disguise or any other media server, it will crash at some point. You will have to restart, and if you have several servers as we did, you could be looking at upwards of 15 minutes to reboot the whole system. The balance of what we were gaining on stage from running real-time content versus the impact on our system needed constant consideration. It is easy to think real-time content is the silver bullet and you will never need to wait for a computer again, but this is not the case. A computer can only be taken so far before it falls over.

To balance the impact on our system and where the rules allow, we rendered a lot of content out of Notch. And we rendered it so quickly! Not always in real time but compared to packaging it up, sending it to a render farm in a different time zone, waiting for it to render, downloading on the venue's dodgy internet connection only to find a randomly flickering light and starting the process again, it was like moving at light speed. And once you have everything in your project turned on, if it doesn't play in 100% real time, it was still doing far better than After Effects ever did. Even if we had to wait 15 minutes for a 5 minute sequence to render, we were saving large amounts of time. And once a 3D scene was rendered it was ready for the stage, it didn't need a final pass through After Effects to polish it up. It was a small 3D revolution. See Figure 8.3 for an example of render time versus result.

This is not to say adapting our workflow was flawless with no moments of frustration. This was a very steep learning curve for the studio. Forty years of collective Adobe experience and logic needed to be unlearned and new logics ingrained in our brains. For the 3D workflow, it was a marked improvement, but for 2D and 2.5D it was more challenging as you are working in an environment designed for 3D. Trying to create 2D composited sequences could feel like you were trying to push the software to do something it didn't want to do. You end up with workarounds and would occasionally find yourself wondering if it would just be easier to fire up After Effects.

Simple and instinctive 2D tasks such as drawing a mask and applying a stroke effect to get a subtly glowing line was and still is very difficult in Notch, but in After Effects it is a 2 minute job. The show has a large circuit board like structure that bursts through the proscenium and fills the auditorium. This scenic piece is made of pixel tape and we animate it a great deal. Doing this work in Notch was proving too time consuming and we were unable to get the precision we needed so we had to switch to After Effects for this. Notch is also

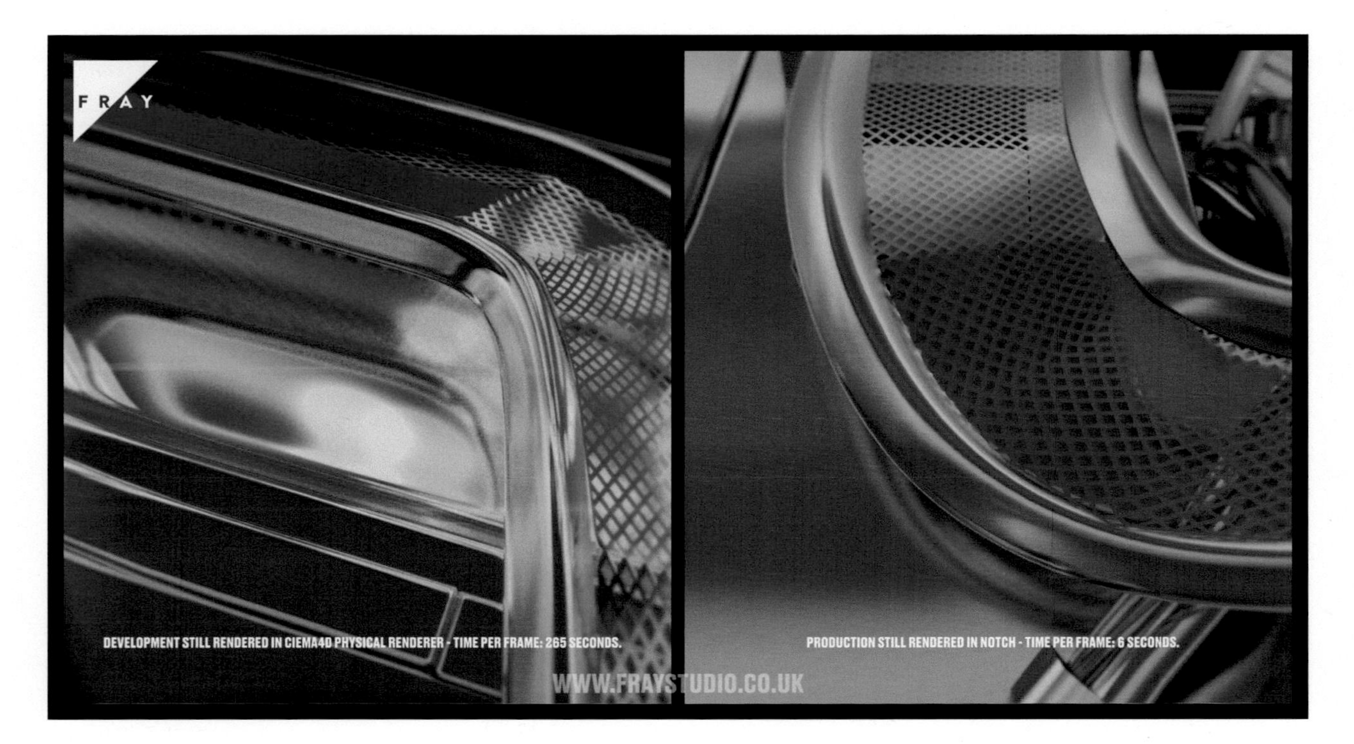

Figure 8.3 Side by side comparison of content production software results. Source: Image Credit: FRAY Studio, 2020.

not great at editing a sequence together in terms of laying out a series of pre-compositions on a timeline. The timeline is not as clean and intuitive as After Effects but After Effects has had 20+ years to perfect this so we could forgive the shortcomings.

Did the major advantages outweigh all the frustrations? Yes. We had so many heavy 3D sequences of the Delorian going back and forth in time that changed almost daily, keeping up with this in a traditional Cinema4D, Render Farm, After Effects workflow would have been frustrating. Notch gave us near instant preview and fast render times that re-gained hours of creatively productive time and allowed new ideas to be actioned quickly and painlessly.

Changing our workflow on this project permanently turned a light bulb on for us. Yes, Notch is not perfect, but nor were the tools we are all used to. In the end, we landed at around 75–80% of the show being made within Notch. Lack of long-term experience and familiarity partly accounts for our falling short of 100% – but there are also some limitations we just couldn't overcome such as mask tracing for the circuit boards. But we are ok with this, it was an experiment that had very positive results. We achieved far more than we had hoped and genuinely felt liberated from the old way of working. We took a significant step toward a faster and more intuitive workflow based around allowing creativity to come to the forefront and not be held back by technology. See Figure 8.4 for another example of an interface for a Notch block used on *Back to the Future the Musical*.

Since 2018 our work has shifted further and further into real-time workflows. Our core tools are now Notch and Unreal Engine, both have their advantages and disadvantages and we use them for very different projects. Fundamentally both allow us to move at speeds previously not possible, we can build ideas quickly and remove the waiting from the creative process to allow more developed work to make it to the stage the first time round. This shift in workflow has also allowed us to move into other areas of work such as broadcast and interactive installation which has enriched the mix of experience we gain as a studio.

It is a brave new world of real-time workflows, and while it has come leaps and bounds in recent times, it is far from a finished project. There is tremendous promise there, and as a creative community, we need to embrace these tools to help develop them together to make software truly fitting for the live-video community.

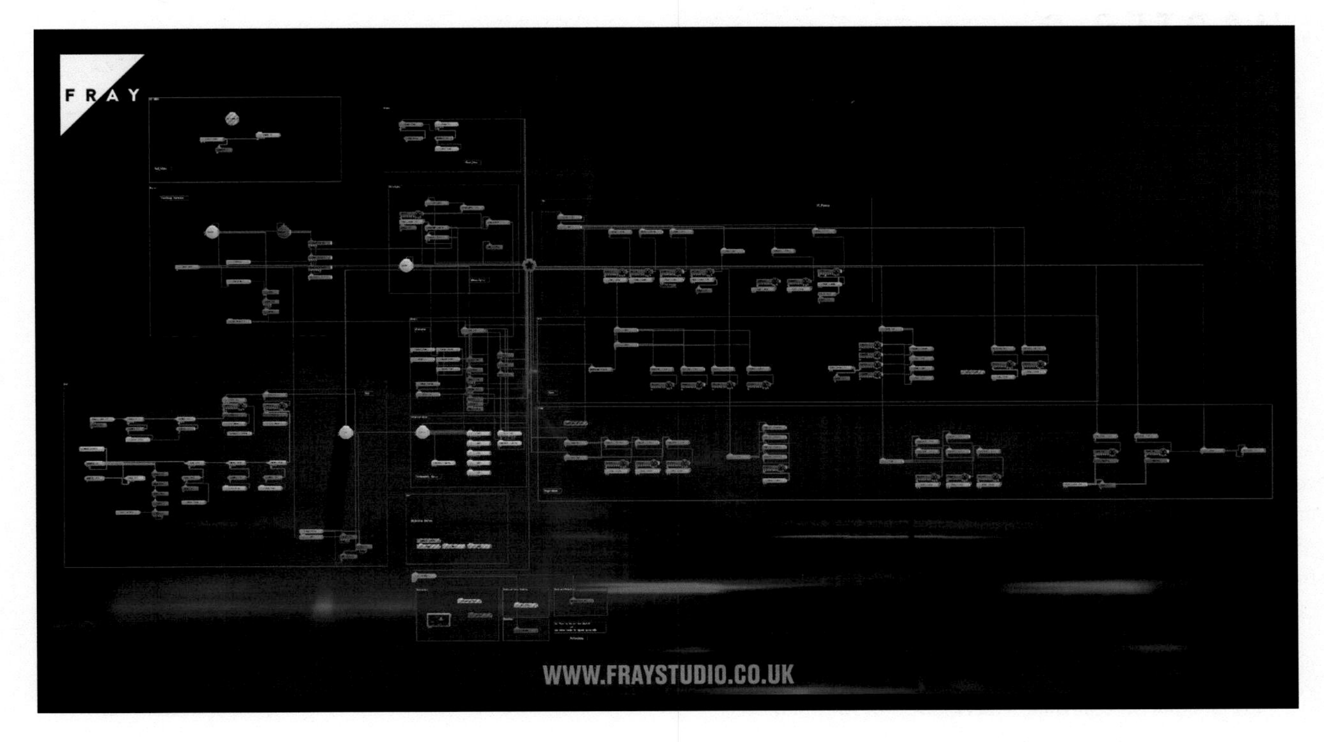

Figure 8.4 Example of Notch node graph for *Back to the Future the Musical*. Source: Image Credit: FRAY Studio, 2020.

by Finn Ross, FRAY Studio
Website: www.fraystudio.co.uk

Fray Studio is an Olivier and Tony Award-winning design studio for the entertainment industry. As leaders in our field, we have video in all kinds of three-dimensional spaces to incredible effect and acclaim. Collaboration, emerging technologies, and creativity are at the heart of our work. We take a cross-cultural approach united by our focus on using video as a storytelling tool; no matter what medium we are working in, the duration of the experience or the technology used.

Project: *Back to the Future the Musical*
Location: West End, London
Year: 2021

Team
System Design: Ammonite – Jonathon Lyle
Programmer: Emily Malone (Manchester) and Neil McDowel Smith (London)
Video Engineer: James Craxton (Manchester) and Sam Jeffs (London)
Principal Animators: Adam Young, Norvydas Genys
Animation and Notch: Henrique Ghersi
Animation Development: Laura Perret
Video Assistant: Kira O'Brien
Video Technicians: Matt Somerville
Video No1: Ollie Hancock
Video No2: Piers Illing
Client: InTheatre Productions

CHAPTER 9
Augmented Reality

Augmented Reality is an excellent use case for real-time content creation. AR takes a live camera feed and augments the image with computer generated visuals. As the camera moves, AR objects appear as if they exist naturally in the camera shot. In reality, these objects are visible only in the composite of the camera shot and the generated AR objects.

AR elements are rendered to camera POV and composited to the live camera signal in real time. Practitioners of AR workflows typically refer to these generated visuals as a "front plate" as they are placed in the scene's foreground. Advanced AR platforms enable clever ways of intertwining motion graphics with existing scene elements, creating the illusion of depth.

What is the difference between AR and VFX? Augmented Reality is the live visual augmentation of a camera signal and is a subset of VFX. VFX is an image creation and compositing process that historically occurs in post-production, after principal photography is complete. AR processing is a VFX applied to a live camera feed in real time.

AR has seen great success in the game market. Most everyone reading this will own a smartphone that contains all the tools needed to enable AR gaming experiences. The phone's camera, screen, and internal gyroscope for spatial positioning, provide all the data programmers need in order to "place" 3D virtual objects into the phone camera's field of view. Object position and orientation are maintained as the viewer changes camera position. Pokémon Go is perhaps the best known example of game AR. The company behind Pokémon Go, Niantic, built an AR game engine that you can learn more about on the companion web page for this chapter.

Another mobile AR tool you can experiment with right now is the Sketchfab app. This mobile application will give you access to a large online library of 3D resources that you can locate virtually on a physical plane, like a desktop, in your phone's field of view. After placing a virtual object, move around it to see how it maintains its orientation. Sketchfab renders object 3D data in real time using your phone position as determined by the internal gyroscope, then composites it to an internal live camera feed from your phone camera to your phone screen.

> Download the Sketchfab app and try this yourself. Links can be found at rtv-book.com/chapter9

The base components used by your phone to create an AR experience are analogous to those used in a broadcast studio or performance stage: camera, screen, and position data. In phone app use cases, what the camera sees is combined with information about camera orientation to include newly created visuals with the same orientation and deliver the result to the screen. In a studio or stage, a 3D model of the working environment integrates these components with information about the venue and set (see Figure 9.1). While some phones can generate a 3D model of the surrounding space using onboard LiDAR, not every broadcast AR platform can generate a model. Make sure you understand where the model data for your project will come from and how you can correct your model onsite, even if the model is only the studio floors and walls. Camera data is equally critical, even if the camera is on sticks and is restricted to pan and zoom.

Broadcast AR has a variety of uses. One of the longest used forms of Broadcast AR is for sports graphics, as discussed in the sports section of Chapter 5. Sports commentary led the industry in production of virtual graphic overlays to communicate game stats. These real-time data displays integrate with physical locations on the game field or in the production studio. We see similar applications in weather reporting and other informational storytelling. See the companion website for examples.

Live event broadcasts use AR to enable scenic augmentation, creative video content, data displays, and virtual performers.

DOI: 10.4324/9781003206491-11

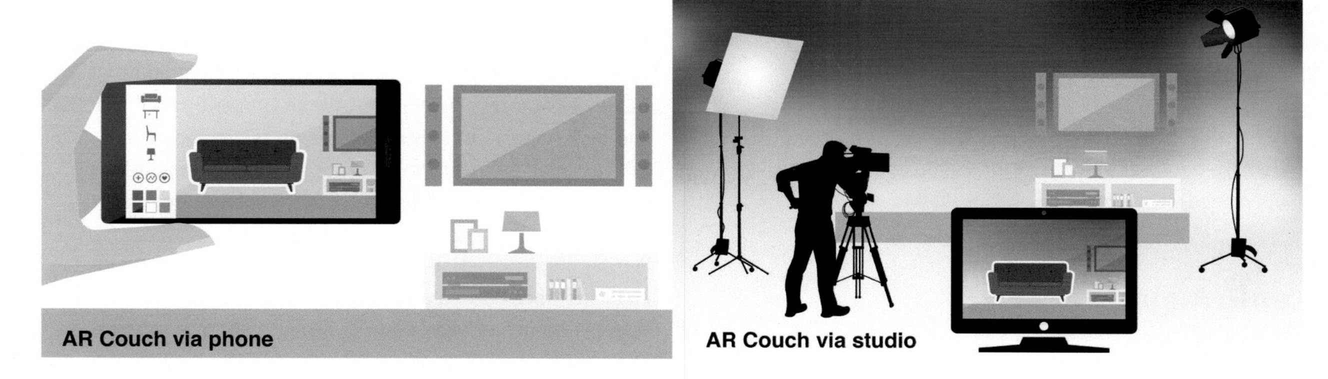

Figure 9.1 Comparison of phone app AR (left) to studio AR (right). Source: Image Credit: iStock.com/chipstudio, iStock.com/elenabs and author.

Using scenic augmentation sets can disappear under a motion graphic snowbank that appears to fill the venue for one scene and then burst into flames in the next. Need an animated character to perform on your live broadcast? A performer can be volumetrically captured and become a cartoon space alien that appears to be on stage. These augmentations become visible through a camera, camera telemetry tracking, and AR compositing.

AR imagery can also track performers or props. Infrared trackers locate an object in 3D space to generate real-time content based on camera perspective and object position. Depth sensing cameras, like Intel's RealSense, enable skeletal tracking of performers and depth measurements of objects in the scene. With this data, imagery can follow a human shape or around a physical object, disappearing when obscured by that object relative to the camera lens. Motion Capture suits collect this data even more precisely and are used for real-time digital puppetry.

There are forms of AR that don't need to rely on camera position data, but work off other image tracking technologies instead. Most current video chat software programs contain tools to track facial features and augment a face with full eyebrows, freckles, or a full cartoon head. Machine learning algorithms adapt augmentations to the details of your face without any depth or model information.

In these examples, systems generate images in real time based on the position of the camera or interpretation of the camera signal. This process occurs in milliseconds so that the AR visuals appear in sync with the live camera footage.

Think about that for a second. A frame of video represents 1/24th to 1/30th of a second. Camera signal, camera telemetry, and 3D data for the scene process practically concurrently to produce a corresponding 3D graphic image aligned to appear naturally in each frame of video. Composite results appear as a seamless video signal within milliseconds.

Let's look deeper at some of the different applications of real-time content creation used for AR.

Virtual Graphic Imagery

In Chapter 8, we discussed entertainment based uses of real-time content creation. For these examples, we could assume that the real-time creative video content appeared on a screen as part of an installation or stage design.

When we combine real-time creative video production with AR tools, we give content designers the opportunity to take content off the screen, putting visuals anywhere in space, unlimited by the boundaries of the screen edge. With AR, we augment the reality a camera captures before delivering it to a viewer. Video design for live entertainment can exist anywhere in the camera frame, but is dependent on that camera. AR cannot occur unless the live image is captured, processed, and delivered to the viewer through a screen: an IMAG screen, a television, a computer screen, a phone screen, or an AR headset.

Concert tours and television broadcasts have used AR creative video as part of their production design. Real-time generated images may support content in the video screens in the set or may be used independently outside the boundary of the video screens with the aid of a camera. Content in the screens can flow seamlessly from screen to air with AR. The magic happens when a camera move exposes the depth of the content, suspended in air or emanating from the floor of a set. Even the simplest particle systems and flat ribbons come to life.

Visit rtv-book.com/chapter9 for examples of AR creative video used as screen independent motion graphics.

Scenic Augmentation

For entertainment environments, Scenic Augmentation is an ideal use of AR. This can take many forms. For example, on a small stage, an AR set can extend beyond the physical boundaries of the stage. The audience sees a performer in front of the physically built set, but in a wide shot, the set contains 3D virtual elements only visible on camera, added with AR processing (see Figure 9.2).

Going a step further, the entire stage may consist of a video environment made of LED screen. The edge of the LED wall might be visible on camera like our example of a physical set. At the edge of the screen, a virtual set extension can extend the image content off the video wall and into AR compositing. We typically think of AR as part of the shot foreground, but AR is often employed to complete a background image (see Figure 9.3).

Set extension technology requires careful calibration of the working environment. Camera lensing, shutter angle, screen angles, color correction, and shutter speed must be carefully managed to keep the real world captured in camera and the AR world perfectly in sync. Otherwise, the extension point where the physical meets the virtual will not properly stitch together and look fragmented. Scenic extension technology could be a full textbook by itself, however its complexity is immensely valuable for the flexibility necessary to shoot a live event.

Motion graphic overlays designed to enhance the set with elements both natural and magical can also virtually augment physical scenery. Need to turn a beach set into a winter scene instantaneously? That might be a job for AR. Want to have a dinosaur break through the ceiling of the venue? Your budget might be best served by doing that one virtually.

Of course, we see examples like this in film and television all the time, but typically productions assemble these in post-production, well after the initial camera shoot. AR's value is the production of quality generated visuals for live broadcast. While real-time can't yet achieve the photographic quality of top VFX production, real-time can currently produce game quality graphics. Image quality will continue to improve as processing hardware advances.

Human Tracking and Motion Capture

Another entertainment application for AR uses position information of a performer on a stage to track their location

Figure 9.2 Scene captured on a stage: the left half shows live audience view, the right half shows AR augmentation of stage with use of camera and AR compositing workflow. Source: Image Credit: iStock.com/klyaksun and author.

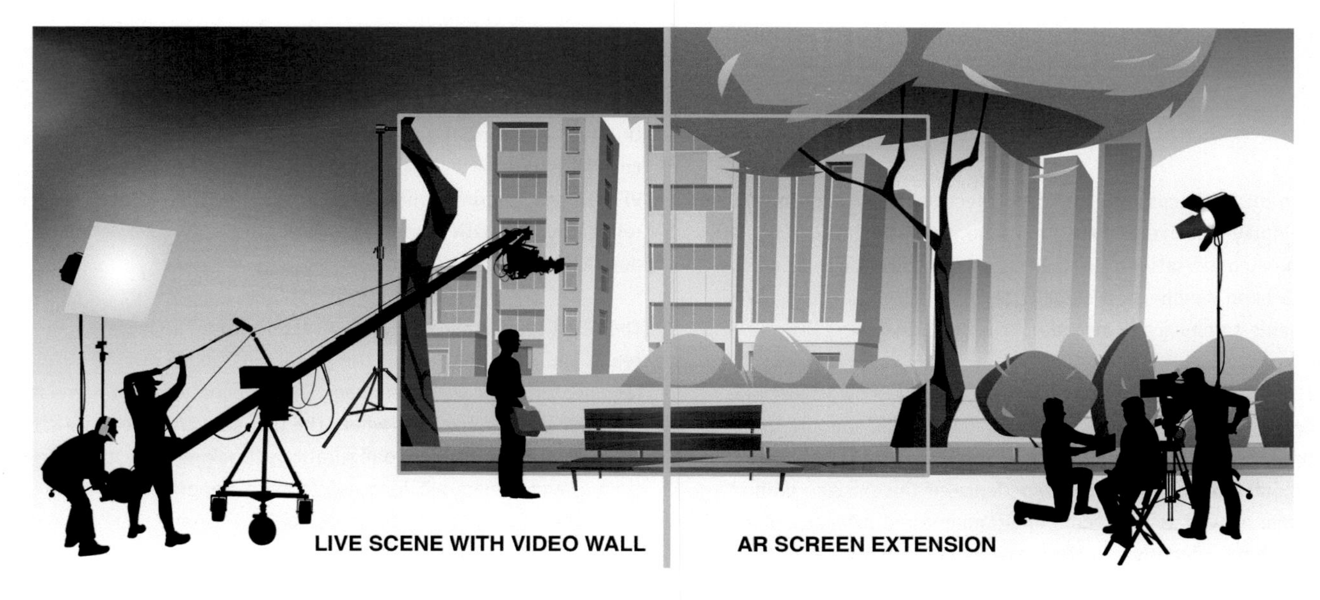

Figure 9.3 Studio with LED wall: the left half shows the live view, the right half shows the scenic extension of video content composited to visible content on LED wall. Source: Image Credit: iStock.com/chipstudio, iStock.com/AskinTulayOver, iStock.com/klyaksun and author.

and actions with motion graphics. We can track people in a space using a variety of technologies: LiDAR (for skeleton tracking), IR tracking, motion capture, volumetric capture, and machine learning. All of these methods provide data for interactive visuals based on human location and action.

For example, let's say a singer wants sparkles to emanate from their hand gestures during their performance. We can achieve that with IR trackers placed in the performer's hands to give position data. Additional IR receivers placed around the physical stage are mirrored in a virtual 3D model of the stage environment to properly locate the trackers. With this information and the camera orientation data, we can generate content based on the location of the performer's hand, locating the origination point of the stars exactly for the camera shot. As the hands move, so too will the origination point of the animated sparkles.

Using a Motion Capture (MoCap) suit, we can turn a performer into an animated character. The suit provides a fast way to cover a human form in position trackers that deliver skeleton tracking data. These data points correspond to animated character rigging points that generate 3D animation in real time. Combine this with an AR production tool set and you can have a shaggy dog perform with your pop star for a live broadcast.

Performers can have their entire likeness recorded in 3D using volumetric capture. This technology creates an animated 3D mesh, textured with a video of the performer. We can then view (and manipulate) the captured data in any 3D modeling environment. In the following case study, we'll see how the team at All Of It Now used a set of 3D models in an AR performance with a live band.

Case Study – All of It Now – Coldplay x BTS

In December 2021, Coldplay and BTS performed "My Universe" on NBC's *The Voice*. This performance collaboration was unique as Coldplay appeared in person and the BTS band members performed as holograms in Augmented Reality. Each BTS band member appeared as 3D virtual avatars created via volumetric capture, and rendered live using Unreal Engine 4.27.1.

> Visit rtv-book.com/chapter9 to see a YouTube clip of the performance.

Project Timeline

The Coldplay team had recently shot a music video for "My Universe," in which the team worked with Dimension Studio to volumetrically capture all seven BTS band members, as well as Coldplay themselves.

Visit rtv-book.com/chapter9 to read an article describing this process.

The initial discussions for this project began in August between Coldplay Creative, Dimension Studio, and All of it Now (AOIN), where conversations involved using real-time AR to bring the BTS band members on stage with Coldplay, even if they were unable to physically attend.

From there, conversations pivoted to find a show where all teams would have sufficient time, access, and support to produce this effect. *The Voice* opportunity was selected by the Coldplay team, due to the experience of the existing production team, as well as the studio environment and infrastructure, which significantly expedited the install process.

Technical Considerations

The volumetric capture assets of BTS with holographic treatment were provided by Dimension Studio, in two formats, an Alembic sequence, and a special mp4 export using Microsoft's SVF Plugin for Unreal Engine 4.27.1.

Due to the high data rate of playing back seven Alembic sequences simultaneously, the mp4 approach was selected instead as the preferred method, and both AOIN and Dimension teams moved forward with creative effects and sequencing.

As with any groundbreaking innovation there were some technical challenges to overcome with the mp4 approach – the current SVF plugin for UE4 did not have the ability to track the mp4 recordings properly to timecode. This required collaboration between AOIN and Microsoft to rewrite elements of the SVF plugin code so that the BTS performers remained in sync with Coldplay on stage.

Another challenge with using the original volumetric capture recordings was that the original BTS performances were recorded at 24 FPS to match the music video frame rate, but *The Voice* is produced and broadcast at 29.97 FPS. This created some lip sync issues, due to frame blending 24 FPS into a 29.97 output, but AOIN was able to clean up the lip sync issues in post-production.

Creative Pipeline – The Creative Process

The creative pipeline involved reviews and discussions with both the Coldplay and Dimension teams. Some visual effects were created by the Dimension team, and then some effects created by the AOIN team, with all efforts merged using Gitlab as the version control platform.

Due to the nature of performer mesh textures, the mp4 format created issues with some initial approaches to the hologram VFX, so the hologram glitch and on/off transition effects were refined by the AOIN technical team, with creative sequencing done inside the UE4 sequencer.

The AOIN team received 3D stage assets from *The Voice* team, and were able to put the performers inside of a 3D representation of *The Voice* stage. This process enabled a crucial round of previsualization, where the Coldplay creative team could test AR performer configurations, tracking area, transition timing, and camera blocking before even stepping foot inside the studio.

AOIN was able to ingest the 3D stage assets and production camera plot into Unreal, creating a real-world scale accurate representation of the seven AR cameras, and the performers, so that the Coldplay creative team could visualize which performers would be visible for each moments, which cameras could best capture these moments, and so previsualization renders became the best method to communicate these moments to all parties involved. See Figure 9.4 for an example.

This previsualization time was crucial in making the best use of the limited time onsite, and also helped unify all technical and creative efforts into a shared deliverable with shared understanding.

On Set Deployment/Challenges

In addition to producing this performance, the *Voice* team was actively producing the rest of the season. This meant scheduling and coordinating all aspects of this performance around the existing performance and production schedule of *The Voice*.

The installation of this AR system relied on a few different installation phases: installing tracking stickers in the studio for the camera tracking system, mapping the tracking area, calibrating all seven camera lenses for AR applications, rehearsing and camera blocking the performance with the *Voice* team, and then finally rehearsing the AR performance with Coldplay on stage, all taking place in seven days.

Deploying and calibrating seven AR cameras while in the midst of a fast-paced TV production schedule was no easy task, but with the help and support of the *The Voice* production team, All of it Now was able to coordinate the calibration and

Figure 9.4 Previsualization Render. Source: Image Credit: All of it Now, 2022.

installation of seven tracked cameras inside the studio around the existing production, without significant disruptions to the existing schedule.

The support and supervision provided by the broadcast engineering team throughout the process was crucial to the success of this project (see Figure 9.5). This installation process is typically where most projects go wrong, but the knowledge and experience of our onsite technical partners of the *Voice*'s production schedule, and suggestion to bring in additional cameras is what made this process so successful.

AOIN worked with Stype and their Red Spy Fiber camera tracking system, and Augmented two Jib cameras already used by the production, and then supplemented that camera package with five dedicated AR cameras, which allowed for calibration of the tracked cameras even while the *Voice* was shooting another segment.

One hurdle was that the *Voice* performance was reduced to 3:30, whereas the original volumetric capture performances were recorded using the original 4 minute long run time. This required shortening the performance, without losing any metadata within the encoded files. As the AR holograms were tracking to timecode, the AOIN team were able to edit the incoming timecode to effectively "skip over" the missing section, hiding the skipped timecode section with a transition.

Post-Production Challenges

After three successful takes, the *The Voice* edit team went to work, and was able to coordinate with AOIN on delivering re-renders of select shots, tightening up the lip sync performance of some BTS members, along with adding more glitch effects on certain moments.

AOIN was able to use the tracking data recorded onsite, along with the clean camera plates to re-composite the AR holograms back onto the clean camera plates without requiring traditional post production camera tracking techniques.

A Volumetric Future

The ability to record once, and reuse assets across multiple applications that extends the lifetime of 3D assets beyond their initially intended use is a major milestone achievement for volumetric recordings. This also sets an exciting precedent for use in both Mixed Reality, and metaverse applications.

Danny Firpo, All Of It Now

Project Credits
All Of It Now
https://allofitnow.com/

Figure 9.5 AOIN team working on production site. Source: Image Credit: All of it Now, 2022.

All of it Now is a Mixed Reality production studio, specializing in real-time technologies for live and broadcast entertainment. All of it Now has experience integrating advanced tracking technology into real-time content, which spans across multiple platforms and applications, including XR, In-Camera VFX, and post-production.

Project name: Coldplay X BTS, "My Universe" on *The Voice*
Project location: Los Angeles
Year: 2022
Project team:
Executive Producer – Danny Firpo
Production Manager – Nicole Plaza
Technical Director – Berto Mora
Senior UE4 Technical Artist – Jeffrey Hepburn
Senior AR Engineer – Neil Carman
AR Engineer – Preston Altree
Client: Warner Music
Production Credits:
Coldplay Creative Team
Coldplay Creative Director – Phil Harvey
Production Designer and Co-creative director – Misty Buckley
Head of Visual Content – Sam Seager
Lighting Designer – Sooner Routhier
Project Manager Creative – Grant Draper
Screens Director – Joshua Koffman
Video Designer – Leo Flint
Screen Content created by NorthHouse Creative

Volumetric Capture Studios:
Dimension, London
Executive Producer/Co-Managing Director: Simon Windsor

Sales/ Client Director: Yush Kalia
Head of Production : Adam Smith
Producer/ Technical Director: Sarah Pearn
Senior Lead Volumetric Technical Artist: Adrianna Polcyn
Animator: Ben Crowe
CG Lighting Artist: Marcella Holmes
Jump Studio (SK telecom), Seoul.
VP, Head of Metaverse Company: Jinsoo Jeon
VP, Head of Metaverse Development: Khwan Cho
Executive Producer: Sung Yoon Baek
Business Development: Taekeun Yoon
Producer: Minhyuk Che
Stage Supervisor: Gukchan Lim

The Voice Team
Executive Producer – Amanda Zucker
EIC – Jerry Kaman
Director – Alan Carter
Supervising Editor – Robert M. Malachowski, Jr., ACE
Post Production Supervisor, Associate Director – Jim Sterling
Assistant Editor – Joe Kaczorowski

CHAPTER 10
Background Replacement

Chapter 9 covered the ways AR is used to place generated visuals onto a camera feed foreground; this chapter looks at use cases that fill entire backgrounds with real-time content. Virtual backdrops depend on the same technology previously discussed: camera telemetry, 3D environment modeling, and real-time content creation tools. Background Replacement requires the use of a physical surface upstage of the performers, scenery, and props. That surface can be a video screen or a backdrop of a single field of color. Background Replacement workflows introduce real-time simulated backgrounds through the use of a chroma key compositing process or via a signal sent to the video display.

AR augments the camera capture of a performance with motion graphics placed in the scene foreground. Scenery can be extended or enhanced with virtual overlays. Performers can interact with real-time animated collaborators. These image resources composited on top of the camera signal are known as front plates. Background Replacement affects the back plate as content that occurs behind the primary physical scene content. These virtual backgrounds are either captured live on camera from a physical video screen or keyed into the frame with a green screen.

Background Replacement combined with Augmented Reality brings back plate and front plate together as eXtended Reality (XR) which we will review in Chapter 11. This chapter focuses specifically on the use of the back plate and how it is used for In Camera VFX (ICVFX) on LED stages, as well as green screen studio production.

The development of camera location dependent Background Replacement technologies, with or without AR, gave rise to the term Virtual Production (VP). Virtual production is a real-time VFX defined by the relationship between a physical camera and a virtual camera. The virtual camera exactly mimics the attributes and behavior of a physical camera in a simulated 3D environment, generating perspective sensitive content in real time.

Live generated back plates are used to replace a scenic upstage wall with a simulated dimensional environment that appears within a camera frame like shooting on location. The back plate is made visible within the camera frame either live with a video wall or in composite by shooting against green screen. Background Replacement VP can simulate all the depth of a wide open vista inside a soundstage. Let's look at some uses of Background Replacement.

Scenic Backdrops

Typically, when a scene is shot on a stage for film or television, the background is static. The environment upstage or behind the performers is scenery or even a flat painted backdrop. Traditionally, production designers created a sense of dimension or depth with forced perspective and other tricks of set design. Lighting amplifies these effects to highlight parts of the background and separate the performers from their surroundings and direct audience focus on the performance.

While we have many ways to create the illusion of depth where the actual physical depth is missing, some scenes call for truly epic backgrounds. Mountain scenes, desert landscapes, ocean sunsets; painted backdrops can't convincingly replicate these landscapes.

However, a physical flat wall backdrop may be the only option. Shooting on location can be a complex and expensive undertaking or even impossible to achieve if the desired result is a background from another planet set in the future. To realize such extreme backgrounds, films shoot in a monochromatic studio environment, commonly painted the color green. VFX departments key out the green and replace it with CGI, Computer Generated Imagery, in the post-production process.

DOI: 10.4324/9781003206491-12

Green screen technology is so accessible that it has made internet stars out of cats showing up in old movies. Check out OwlKitty for a green screen explainer at rtv-book/chapter10

Today, a massive VFX industry dominates film production to serve CGI processing. While we touched on a little history in Section 1, what we see today are CGI post-production tools that drive entire production timelines. Prior to shooting, production teams use previsualization 3D model environments to design camera shots. During shooting, those previz models become real-time Background Replacement. In post, that model will combine with camera tracking data to refine VFX for the final look. More recently, in live television these steps are applied concurrently for broadcast.

Only since the late 2010s have we seen the growth of modern green screen for the live performance market. As with traditional green screen, the color green is keyed out and replaced with the scenic model instantaneously. Modern green screen is a VP technology using camera telemetry and 3D models to generate the proper image perspective specifically for each camera shot.

During this same time period, we began seeing the use of LED screens to create realistic reflections on scenery and costumes in film production. Panels of LED are strategically placed outside the camera shot for the sole purpose of playing video files to be reflected by windshield glass or an actor's face. One longtime challenge of green screen has been the removal of the color green from reflective surfaces. LED screen use to create natural reflections was a beginning step in the adoption of video as a background in film. In the 2010s, LED screens lacked the resolution for Background Replacement workflows, but that issue has passed.

LED screen technology now provides consistent color across modules and sub 2mm resolutions. Imagery that would be used for VP green screen compositing is the same imagery that feeds directly to the LED walls for fully immersive shooting. On an LED stage, the camera captures the background along with its reflections (ICVFX), with the added benefit of being visible to the performers working in the space. Both VP green screen and LED stages are in use today for productions requiring Background Replacement.

Check out rtv-book.com/chapter10 to follow ILM's Rob Bredow from xLab to Stagecraft.

Let's look at the motivation to use Background Replacement technology and the working requirements for its best application.

Depth Simulation

Natural looking depth is a primary motivation to use Background Replacement techniques. We can layer planes or objects in 3D separated along the z-axis for depth. Figure 10.1 details three objects separated by 20 feet.

Using this example as our backdrop, the image on the left of Figure 10.2 shows a static background while the image on the right shows a dynamic background using an LED screen. They currently look identical.

Pan to the left and you can see the relationship of the three objects remains unchanged in the static background while the shift is visible in the dynamic background. See Figure 10.3 to compare the results.

This is a rudimentary example, but it shows why depth correction in dynamic backgrounds is so desirable. Realistic backgrounds turn soundstages into location shoots without transporting an entire production team to a remote area. Depth simulation captured in camera looks realistic enough to replace

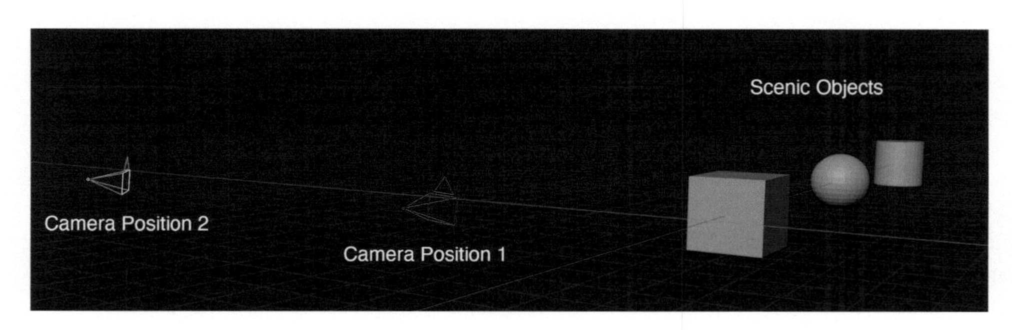

Figure 10.1 Three objects in a line with two cameras and different view angles. Source: Image by Author

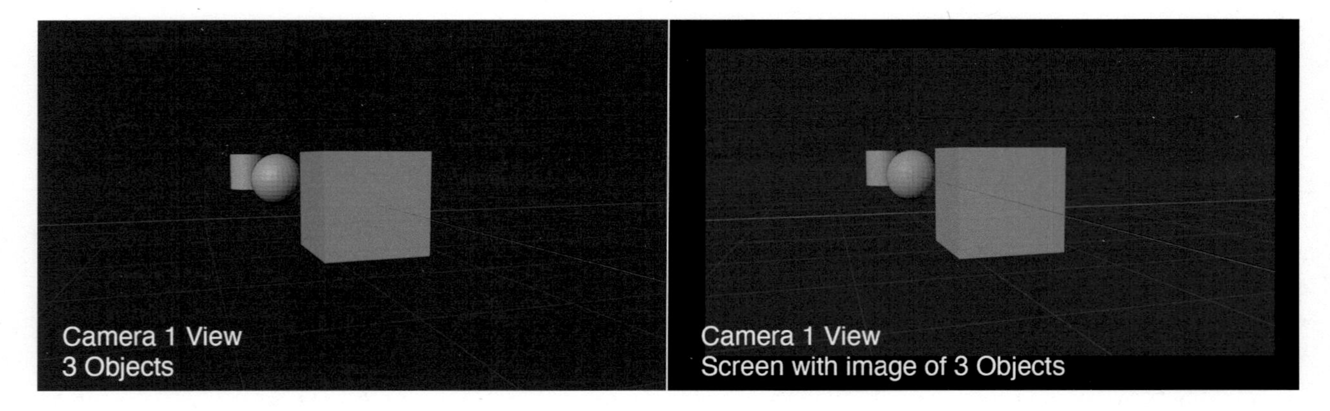

Figure 10.2 Camera 1 View: Image left shows the view of the 3D content, image right is the content on a 2D screen surface. Source: Image by Author.

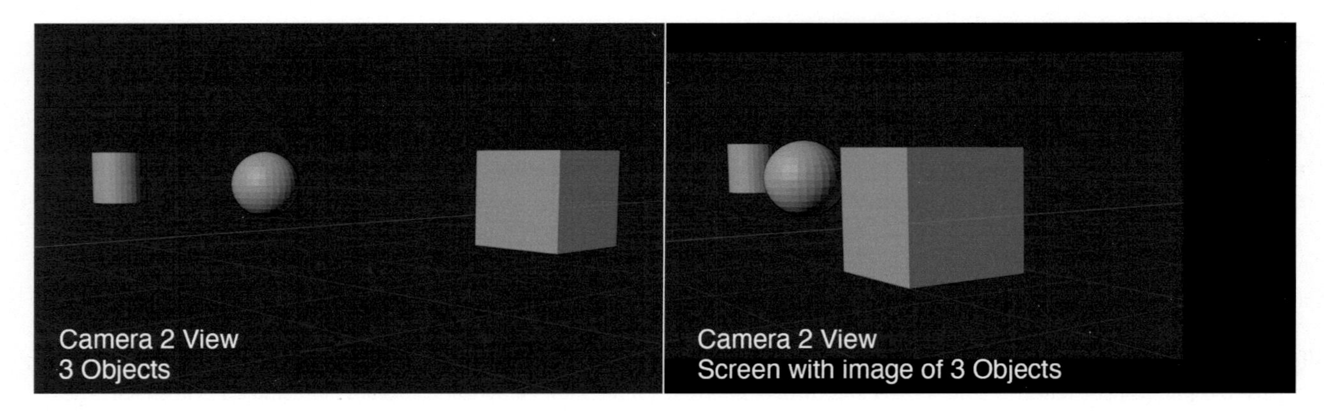

Figure 10.3 Camera 2 View, the camera has now panned left to expose depth: Image left shows the 3D content view while image right shows the flat screen with static content. Source: Image by Author.

location shooting, as Background Replacement technology enables us to precisely replicate the scenic environment relationship to the camera.

While the camera moves in physical space, the data describing camera position, orientation and lens data refreshes hundreds of times per second. The platform used to create background content then duplicates that data to a virtual camera with the exact same specifications as the physical camera. Depth viewed by the virtual camera is used to generate the real-time content used for Background Replacement, giving a sense of depth to the image captured by the physical camera.

High data refresh rates for camera telemetry and image processing are critical for natural looking content. The more frequently a system can update camera position data to the real-time content platform, the more precise and naturally looking the result. The more robust the graphics processing, the more complex imagery can be generated.

By way of analogy, think of a computer game console. The game controller defines what is visible within the game space. A player's actions on the game controller move a virtual camera within the game, and what that camera "sees" is rendered to the game screen. The pan, tilt, and zoom as well as x, y, and z location of a studio camera is analogous to the game controller. Changes in those coordinates reposition a virtual camera in a real-time content platform and what that camera "sees" is rendered to the back plate. See Figure 10.4 for a diagram outlining this relationship.

The back plate generated through the relationship of the virtual camera and physical camera is composited in real time using one of two techniques: live composite using a chroma key of a colored backdrop, or via live feed to video wall backgrounds,

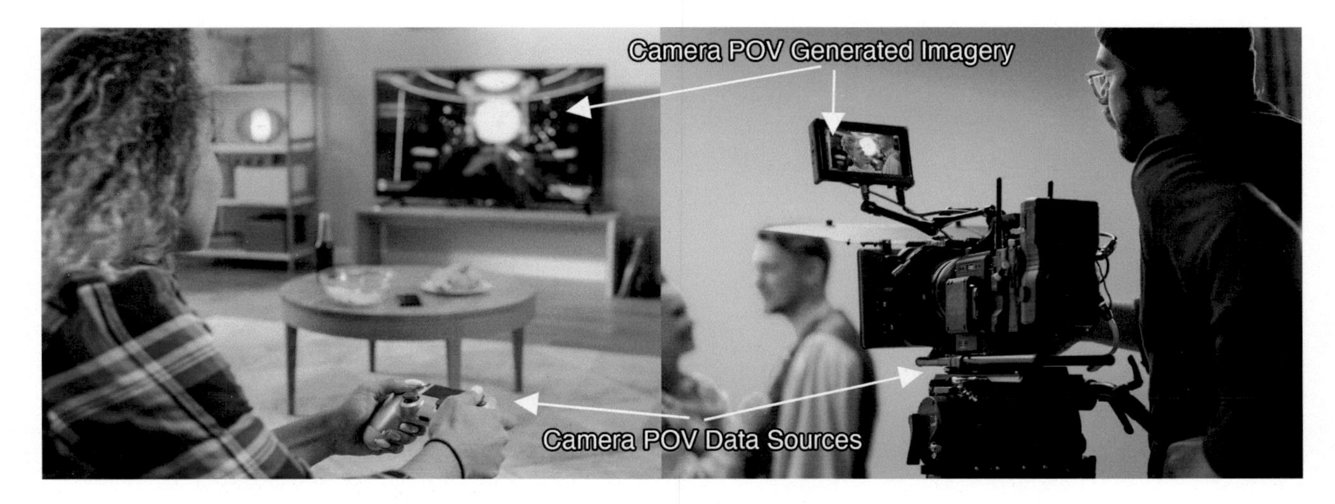

Figure 10.4 Game controllers and screens have analogous counterparts in Virtual Production. Source: Image Credit: iStock.com/gorodenkoff and author.

most often, LED surfaces. Next, we will compare these two Background Replacement production tools.

LED vs Green Screen

Green screen (see Figure 10.5) versus LED, both can replace backgrounds in real time, but how do they differ and why should we consider one over the other? LED stages appear in a variety of configurations composed of LED tiles. Here are several examples: two walls and a floor (see Figure 10.6), three walls with floor (see Figure 10.7), and curved at center with floor (see Figure 10.8). Fully curved LED assemblies might be partially or fully enclosed and utilize a ceiling instead of a floor (see Figure 10.9).

A green screen studio or an LED stage can include floors and/or ceilings. In a green screen studio, green paper, fabric or painted studio walls must be evenly lit to facilitate the keying process. The color green can extend to the floor and sides, surrounding the physical scenery and props wherever back plate imagery will be used. The key color only needs to extend just beyond the boundary of the live captured objects and people in the shot. Once that boundary is established, the generated background can fill out the rest of the frame, easily extending beyond the physical edge of the green backdrop.

Film and live broadcast have different approaches to the use of green screen and real-time back plate creation. In a film style shoot, the green is captured in camera for post processing. Real-time content will be utilized for on set previz of the VFX work to be done in post or upcoming stunt shots, but not for final image creation. Meanwhile, live broadcast green screen is real-time image creation and compositing for the final image destined for immediate transmission.

On an LED stage, it is preferred to fill the entire shot background with LED wall tiles. In a wide shot, this isn't always possible, which requires the back plate imagery to be extended off the physical LED screen using AR technology. The inclusion of LED tiles in the ceiling can be used for lighting and reflection sources, though usually they must be arranged to account for necessary lighting gear. Ceiling and floor LED panels may also appear in a shot as back plate screen output. LED floor panels are valuable for Background Replacement world immersion looks, though challenging to manage for the unnatural lighting angle on faces and props.

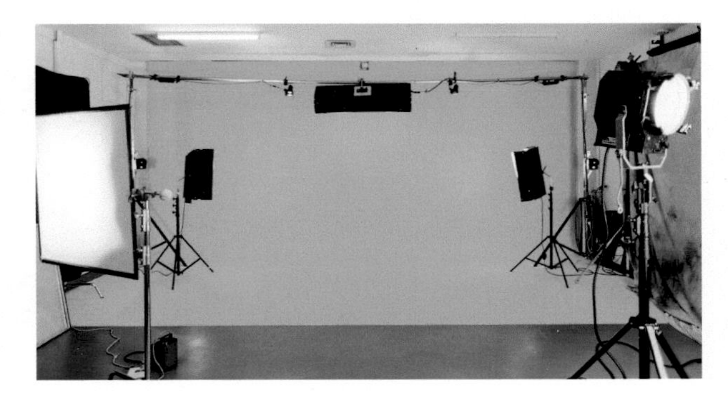

Figure 10.5 Green screen studio. Source: Image Credit: iStock.com/slovegrove and author.

Figure 10.6 Cube style LED stage. Source: Image Credit: Meptik.

Figure 10.7 Three walls with floor LED stage. Source: Image Credit: XR Studios.

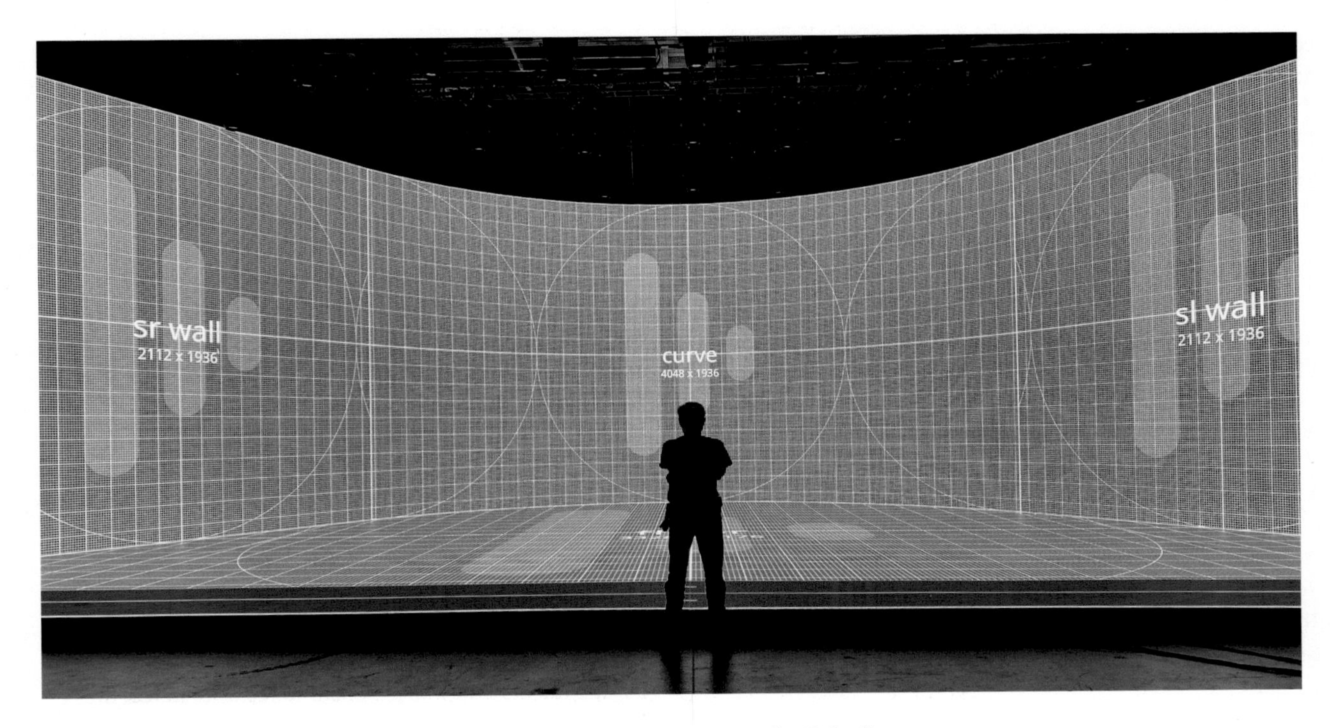

Figure 10.8 Curved center with flat side walls and floor LED stage. Source: Image Credit: XR Studios.

Figure 10.9 Curved wall volume style LED stage with ceiling. Source: Image Credit: Bild Studios.

Budget, production type, and scene design will all impact the choice of VP stage. No single rule tells us what type of stage goes with any particular style of VP production. Film style productions currently favor large curved wall LED stages with ceilings called "volumes." Camera shots are preplanned and choreographed, typically utilizing one or two cameras at a time, favoring looks that maintain the full LED surface within the shot. Real-time content creation dominates the ICVFX targets of perspective generated backgrounds, lighting and surface reflections, however the captured footage is still destined for post. Color correction, potentially other VFX content and other non-real-time processing is still to be applied.

In live broadcast production, we predominantly see smaller LED stages with flat LED walls and floors. Wide shots are common and AR tools are used to extend the virtual scene beyond the LED walls or to add motion graphic elements into the shot foreground. Multiple cameras are typically tracked for XR use. These stages offer a great solution for small studio spaces that can't support a large physical set or need to frequently change the scenic environment with virtual setups. Flat walled LED stages are commonly temporary builds and easier to construct and calibrate than the large curved volumes.

In the next chapter, we will explore some factors that might affect the type of VP stage chosen for a production. First, let's learn about a graduate student project that used Background Replacement for a short film shoot on a flat wall LED stage. This story provides valuable insight for anyone who feels this technology seems out of reach to those working on a modest budget.

Case Study – Fernando González Ortiz – *The Lion and the Firebird*

The New Frontier in Student Filmmaking with Virtual Production

The Lion and the Firebird is the first independently produced short film to be shot using real-time Virtual Production in its entirety. This is the story of how it got made.

Early on during the pandemic, I began doing some research on the state of film, and one of the most recent developments was this "new" technology being used; Virtual Production. By combining real-time rendering with LED walls, you could transport any production to almost any environment imaginable. The real question was, can an independent student production use the technology successfully?

As with all new technologies, this one also came with setbacks. You need high-processing computers, high-fidelity LED walls, genlock-capable cameras, and, lastly, technical skills to use 3D software. While I didn't have access to some of these, I had a computer capable of running a real-time engine. I downloaded it and started watching the tutorials. I went to as many webinars as I could get myself into, and I read as many whitepapers as I could get my hands on. At the end of the day, as a producer, I didn't need to be an operating genius with the technology; I just needed to understand it well enough where I could facilitate and support a group of creatives to work with it (Figure 10.10).

As I studied the technology, I realized that most productions using VP were big-budget productions. In some other instances, the projects were either tech demos or case studies. As with most technologies, the early implementations start as Research & Development, then they get used by a few trailblazers. Later on, they get used by big players, and eventually, a majority can access them. VP wasn't the exception. But what was stopping me from trying?

I set myself in motion. I had to find the right partners for this endeavor. It meant finding technological partners and a director with a script that would benefit from it. At times the early use of technology tends to focus on the capabilities of a new tool and neglect the elements that make a film great. I wanted to focus on a great story. I was on the lookout for a script with a great plot, endearing characters, and a world with prominent visual elements that would be hard to replicate in real life.

Around this time, I met Daniel Byers, the writer/director of *The Lion and the Firebird*. A historical short that explores the relationship between the last neanderthal and a homo sapiens. The short was perfect for VP technology. The director had this vision to present a historically and scientifically accurate story. He knew that VP could allow him to explore a paleolithic world filled with volcanoes, majestic steppes, and mammoth carcasses.

Finding the right story and director was the first step toward making this film. The next was to connect with a mentor and partner to find an LED stage. One of my professors, Maureen Ryan (Columbia University School of the Arts, Professor of Professional Practice, Film), introduced me to Laura Frank. My meeting with Laura was enriching and motivating. She was a very supportive mentor; not only did she answer all my queries, but her guidance demystified the technology. She also connected me with who would become my next mentor and collaborator, Shelly Sabel. I met with her online, and not only did we have an enriching creative conversation, but she also invited me to an LED stage as a learning opportunity.

At the stage, I met the wizards behind and in front of the screen. My practical knowledge of physical production finally coalesced with my theoretical understanding of Virtual Production. That first day I learned more than I could ever have learned by watching tutorials at home. I left that set reinvigorated. I knew that I had to make my short film using this technology. We had to find a way to make it viable for a

Figure 10.10 Producer at work. Source: Image Credit: Matt Infante – Unit Still and BTS Photographer.

small student production to shoot with this costly technology. This initiated a year-long process of securing financing from grants, donations, and investors. With the support of my mentors and a grant by the SLOAN Foundation for our accurate portrayal of science, we were able to get the first pieces of the budget puzzle together.

We looked for talent both for our virtual world and our physical production. We found seasoned virtual artists that guided us, and we had the eager excitement from our physical production crew to explore VP. What we didn't have was an LED stage. I went back to my mentor Shelly to find that her employer had migrated their in-house VP capabilities to a bigger stage. We collaborated to establish an educational partnership. Our production would get a discounted rate, and in return, we shared an opportunity to work in a VP discipline Shelly's team hadn't yet explored: narrative fiction.

Working on an unprecedented project like this meant that the whole team was met with learning opportunities at every corner. The production had a Lead Environment Artist, but there was no 3D artist that could help in the VAD (Virtual Art Department). We had to rely on the Quixel library and the Unreal Engine marketplace. No matter how great our virtual scenery was, we needed to supply the right foreground elements to extend into our physical set. We worked with a resourceful production designer capable of building and outfitting the stage to match up with our virtual environment. See Figure 10.11 for a look at our team at work.

All of this was accomplished with a limited budget and no access to LiDAR scans, 3D printing, or major set design fabrications. Our art department focused on bringing physical elements to the stage that could accentuate the realness of our environment and camouflage the boundary between the screens and the floor.

It wasn't all a seamless experience. While the LED stage crew was familiar with the technology, for most of our crew the shoot was their first time working like this. Time considerations had been built into the scheduling of our shoot, and while our 1st AD had padded our schedule to account for technology

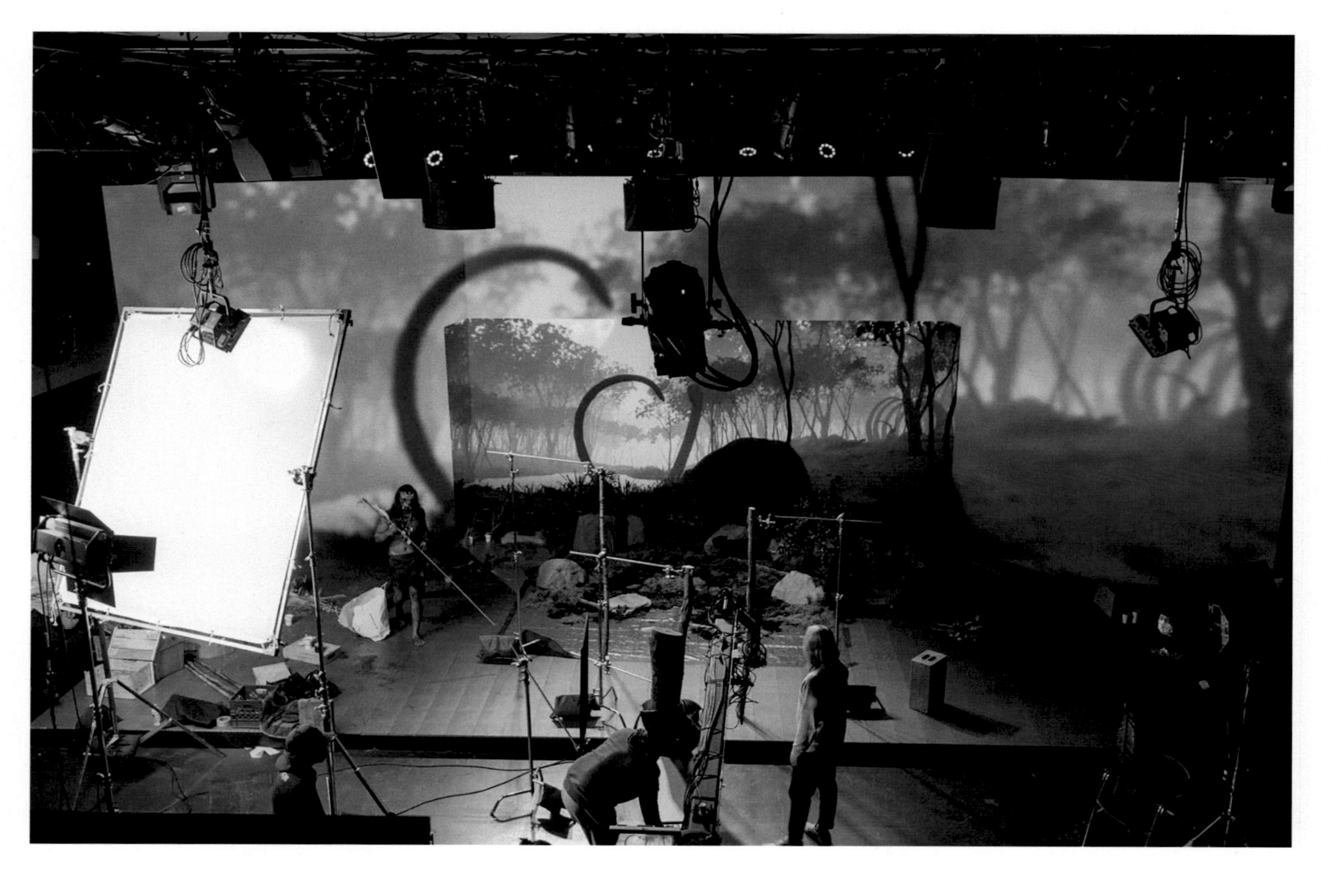

Figure 10.11 LED stage and production team preparing a shot. Source: Image Credit: Matt Infante – Unit Still and BTS Photographer.

integration, it was never enough. Due to the nature of our project, we hadn't had enough time or budget for previz, and our schedule was affected. The director and I had overestimated how much time we could save by eliminating company moves to new shoot locations.

While filming in Virtual Production, we encountered time-consuming challenges such as camera and engine calibration that had to be resolved in creative and technical ways. Since our film required the use of multiple camera support, different camera rigs, and constant changes of lenses, that meant the motion tracking system needed to be calibrated numerous times. Not only did we face time-consuming technical issues with calibration, we had banding and scan lines visible on the LED wall. Regardless of these challenges, our crew was able to keep pushing forward and solve them. These issues could have been anticipated if we had dedicated more time during our pre-production week to the tech scouting. We had a complex shoot with five different environments, and ideally, we would have dedicated at least half a day of calibration and optimization.

Challenges will always arise, but the real question will be how much time we will need to anticipate to solve them. At times, we were forced to stop shooting and calibrate. This was a nightmare for the practical side of it; everyone was ready, but the digital world wasn't cooperating. Sacrifices had to be made to give our director enough time to get the coverage the scenes needed. At times, we had to continue filming, even against the better judgment of our VP team. As VP becomes more ubiquitous, these disruptions will be more effortlessly managed and the tech problems less common.

The reality is that a 100% Virtual Production should not be a priority for any filmmaker. VP is just another tool at our disposal to tell a compelling story. Virtual Production is an illusion that needs to serve the story's purpose. It will be the combination of the artifice of real-life and virtual smoke and mirrors of VP that supports excellent storytelling and leaves people in awe. See Figure 10.12 and Figure 10.13 for stills of the performers in front of their virtual background.

The essence of filmmaking is all in the make-belief of what is put in front of a camera. I believe that Virtual Production will

Figure 10.12 Unit stills for *The Lion and the Firebird*. Source: Image Credit: Matt Infante – Unit Still and BTS Photographer.

Figure 10.13 Unit stills for *The Lion and the Firebird*. Source: Image Credit: Matt Infante – Unit Still and BTS Photographer.

become an important tool used to accomplish the "unobtainable scenes." VP technology allowed us to transform a black box theater space into the paleolithic era from 40,000 BCE. We created majestic caves, expansive forests, fuming volcanoes, and everlasting sunsets. These tools will enable future students to tell new more imaginative stories. Who says that all student productions need to exist within the confines of accessible real-life locations? Student film productions have the opportunity to create new worlds. May this technology allow young filmmakers to dream bigger, just like we were able to do.

The possibility to work with this technology is within reach. It will require sacrifices, self-learning, finding the right partners, and some money, but I can promise that it won't be as complicated or as expensive as it may seem. May more stages and mentors allow young filmmakers to dream big and make unbelievable art.

By Fernando González Ortiz

Project Credits
The Lion and the Firebird
https://darktowerfilms.com/lfb

Project location: New York City
Year: 2022
Project team:
Daniel Byers – Director, Writer
Fernando González Ortiz – Producer
Munir Atalla – Co-Producer
Katrin Redfern – Co-Producer
Francisco Barros – Executive Producer, Lead Environment Designer
Cat Bigney – Executive Producer, Art Director Consultant
Jonathan Betzler – 1st Assistant Director
Zachary Ludescher – Director of Photography
Matt Infante – Unit Still Photographer
Sadra Tehrani – Production Designer
Jack Chen – Virtual Production PA, VFX

Virtual Production team/vendor:
Stage: Former WorldStage LED Volume at The Duke On 42nd Street
Shelly Sabel – Creative Director
Colton Suarez – Project Management, Lightning Board Director
Michael Kohler – Virtual Production Producer
Briana Torres – Virtual Production Integrator

Visit rtv-book/chapter10 for the full production credit list

"Produced in partial fulfillment of the MFA degree in Film at Columbia University School of the Arts."

CHAPTER 11

Extended Reality and Mixed Reality

When you combine real-time Background Replacement with Augmented Reality, you create XR. XR, as used in the entertainment industry, is short for Extended Reality. Mixed Reality occurs when XR or its components are further expanded with interactive information from the real world. Usage of these terms, or "all the Rs" as I've heard it said, gets easily overshadowed with their use in the tech sector. VR, AR, MR, and XR are most commonly used for technology intended for use in the gaming and internet markets. In the entertainment industry, the use of these terms continues to evolve. In the race to develop the metaverse, lines between how these terms are used blur even further.

XR and MR are typically used interchangeably when applied to Virtual Production. However for this discussion, we will use XR and MR to mean closely related but different workflows.

XR combines AR and Background Replacement to complete a live video signal. MR combines AR and/or Background Replacement with interactive content generation from real world events on stage. XR and MR, like AR and Background Replacement, are types of Virtual Production that exist in the larger VFX ecosystem. Like all VP technologies, XR and MR rely on real-time screen + camera + model + content interactivity.

XR and MR are forms of interactive content creation. XR is camera telemetry derived interactive content. The position and lens state of the camera informs production of front plate and back plate video content.

MR also includes interactive content triggers that happen in front of the lens. Examples of MR include IR tracking, lighting interactivity, or any real world performance driven event trigger that generates imagery to the front plate and/or back plate.

This chapter explores how these toolsets are combined and deployed on VP stages. We also review factors to consider when choosing the most appropriate VP stage for a given project. First, let's look at the basics of XR workflows.

Front Plate and Back Plate

To differentiate foreground imagery existing in the AR workflow from background imagery destined for the Background Replacement workflow, we separate real-time generated content into a front plate and a back plate. The compositing process brings these layers together (see Figure 11.1), sandwiching the live captured footage between foreground and background imagery. In the most typical arrangement, we assign background imagery to the back plate, which exists behind any physical scenery or actors, and foreground imagery to the front plate, which exists in front of the set and performers. Later in this chapter we will look at a few of the special cases that don't follow this exact structure.

In our first example, we have a scene of a fire pit in front of a forest scene. The forest is real-time rendered from a 3D model utilizing a game engine. The fire pit, rocks, and performer are live in a production studio. Meanwhile, the flames and glow of the fire are generated through a separate real-time content creation tool. Figure 11.2 shows the final composite.

Looking at Figure 11.1 again, we see the forest assigned to the back plate, and the fire assigned to the front plate. For content creation, we render front plate and back plate imagery separately, even when generated by the same content creation application. In our example, let's assume that we will shoot our scene on an LED stage and render the back plate in real time via a game engine. The resulting video signal is delivered to the LED screens. The fire is real-time rendered via Notch and composited into the camera signal which has captured the live

DOI: 10.4324/9781003206491-13

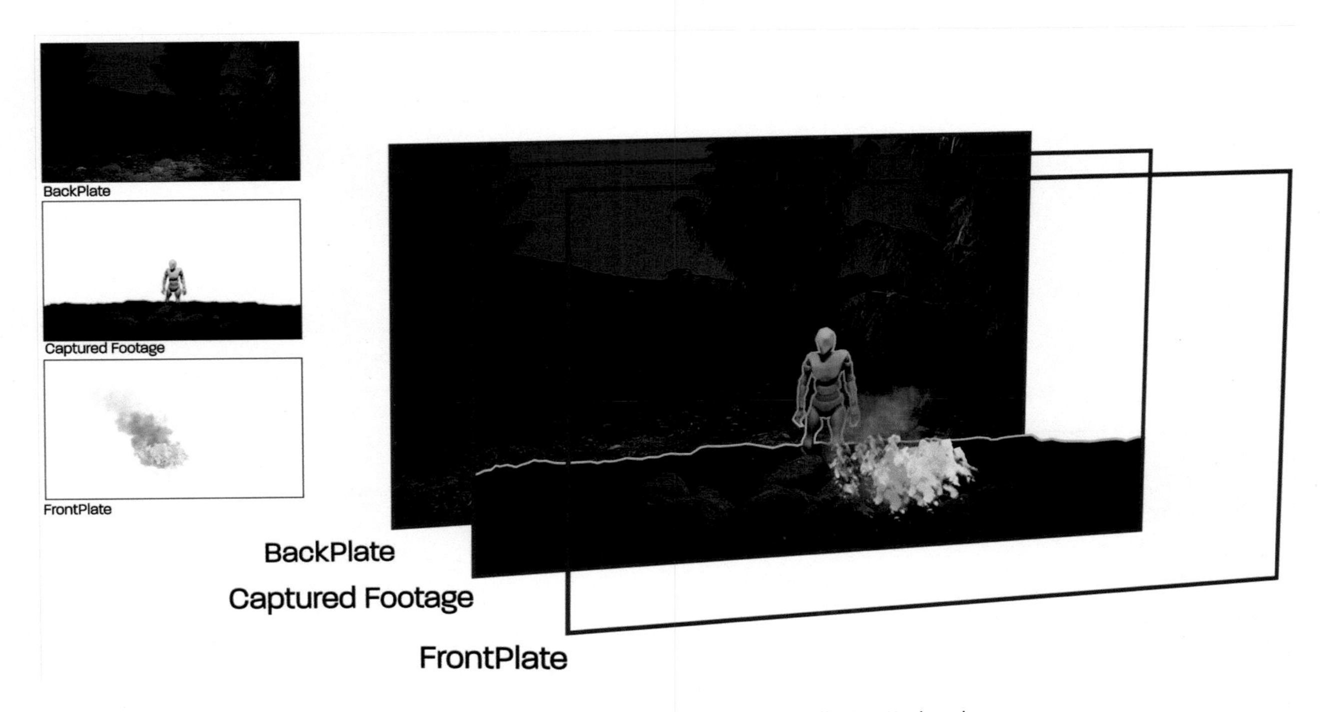

Figure 11.1 Layer stack of front plate, captured footage and back plate. Source: Image Credit: Ian Macintosh.

Figure 11.2 Composite of layered imagery outlined in Figure 11.1. Source: Image Credit: Ian Macintosh.

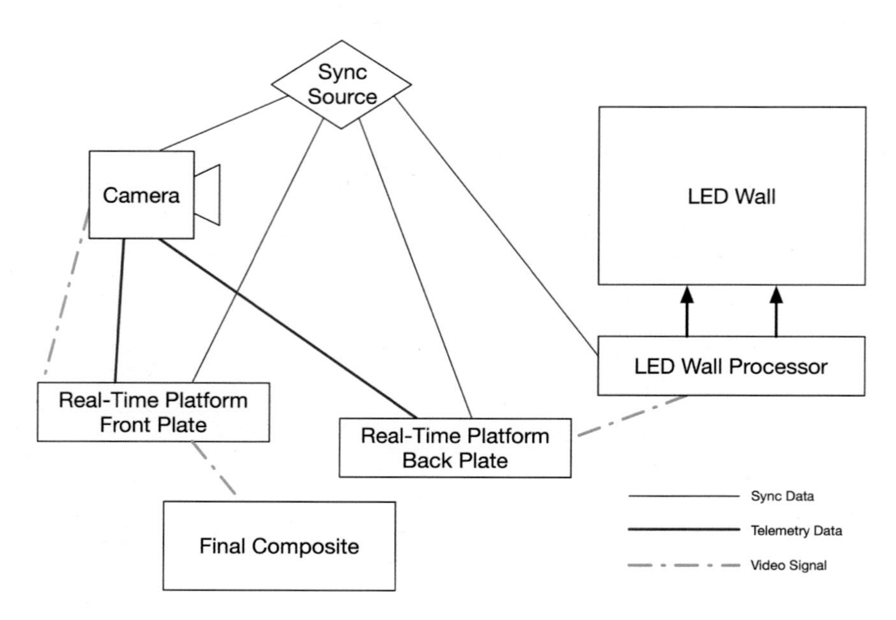

Figure 11.3 Basic diagram of data and video links between camera, real-time platform, and sync source. Source: Image by Author.

background of the forest. The camera signal is a middle layer in the composition of background and foreground elements. Camera telemetry is delivered to the platform hosting the rendering pipelines along with the camera signal.

For this mix of signals to result in a clean live visual effect, we must keep the systems generating imagery for the different plates in perfect sync. All the devices in the system must synchronize with or "listen" to a single master clock in order to draw the correct image state to the intended frame of video at the right time. To accomplish this, a signal from a sync source, like genlock, is sent simultaneously to the real-time platform, the cameras, and the LED wall processor. Figure 11.3 depicts a signal map of the XR process for an LED stage.

For more insight into genlock and VP, visit rtv-book.com/chapter11

Each point of signal transmission and processing can add a frame or two of delay. At 30 frames per second, five points of signal processing could cost a third of a second of delay. This is something that can easily look like a glitch to a viewer if not properly managed. Using our example in Figure 11.2, if the back plate generated for the LED wall takes a frame to process the location of the camera and generate the corresponding background, we must account for that amount of delay in the front plate processing. We must also adjust timing to compensate for additional switching challenges when multiple cameras are present for a broadcast transmission. Not only do the cameras and compositing platforms need to be in sync, the correct back plate must be delivered to the LED stage on the exact frame the corresponding camera is made active. Managing successful delay structures should be a clear and thoughtful discussion between the XR and Broadcast Engineering teams.

Front plates and back plates must be rendered for each camera in the XR signal pipeline. In a live broadcast scenario with three XR active cameras, there are two approaches: isolate every plate onto a unique render host and deliver the composite to the broadcast switcher, or employ an XR platform cued from a broadcast switcher that only renders the front and back plates for the active camera. Again, these decisions are best made with discussions including production members outside the XR team. Budget, schedule, experience, and expertise will inform the ideal approach for each production. While there is critical expertise within the XR team, this work is not isolated to the XR team, and requires the best capabilities of broadcast engineering, camera direction, and producing to succeed.

Sync

Sync is critical to the success of these scenarios. When assembling a system, we must understand each step in the signal transmission and image processing in order to set up coordinated delays in the signal paths that keep real-time content aligned to time. This includes structuring your real-time platform to render consistently for simple or complex files.

For example, if the AR host platform renders a complex 3D model for one scene, and a simple 2D plane for the next, we can't allow the system to deliver the more easily processed frames faster or the entire structure will fall out of sync. Rendering times must be consistent. More often, complex scenes need to be simplified to keep real-time real.

Each production build will have unique sync challenges. Overcoming these challenges requires engineering team input from the camera, screen engineering, and broadcast departments. Once these teams identify and establish a system for managing the delay structure, the front plate and back plate will appear perfectly composited to the live scene. Sync is one of the seemingly effortless behind-the-scenes bits of magic that is in practice an invisible beast of complexity.

Besides sync issues, another complexity of layering of front and back plates with the live scene occurs when back and front plate imagery change order position in the composite layer structure. There are occasions when the virtual background is part of the AR workflow. Sometimes front plate imagery is obscured with physical scenery or performers. This represents a dynamic change from foreground to background (and possibly back again). Let's look at how we can manage these special cases.

Scenic Extension

In Chapter 9, we discussed the use of Scenic Augmentation as a type of AR processing. Scenic Extension is a special case of Scenic Augmentation. Back plate imagery in use on a physical LED screen in a Background Replacement workflow is made visible when the edge of the LED screen is on camera. This typically occurs in a wide shot or when multiple cameras are in use. The back plate must flow seamlessly from the edge of the LED screen into the front plate of the AR workflow. When this composite succeeds, the background appears continuous from the real world of the LED wall to the virtual world of the AR composite process.

Using our campfire scene example from earlier in this chapter, let's imagine a camera shot in which the edges of our small LED stage configuration are visible on camera. Our goal is to continue the background and make the forest scene fill the entire frame, similar to what is shown in Figure 11.4. The image on the left shows the stage with our back plate contained within the boundaries of the LED walls. How do we get the rest of the background visible as shown on the right of Figure 11.4?

In this shot, we need to extend the background content present in the LED screen into the rest of the raster area. Three processes are at work: 1) the AR fire renders as a front plate, 2) the section of background forest destined for the LED wall renders to a back plate, and 3) the section of background forest visible to the camera, but located off the LED screen, renders as another front plate. Figure 11.5 provides a visual outline of the layer structure.

As the camera shot moves or changes zoom, the relationship between the background delivered to the back plate and the background delivered to the front plate must remain aligned to appear seamless (see Figure 11.6). This is a very challenging sync management and camera calibration problem. When it works, the result is the power of Virtual Production on a small LED stage format.

Figure 11.4 Image on the left shows the LED stage with a back plate. The image on the right shows the same stage composited with a Scenic Extension using AR. In a good scenic extension workflow, physical and virtual perfectly align. Source: Image Credit: Ian Macintosh.

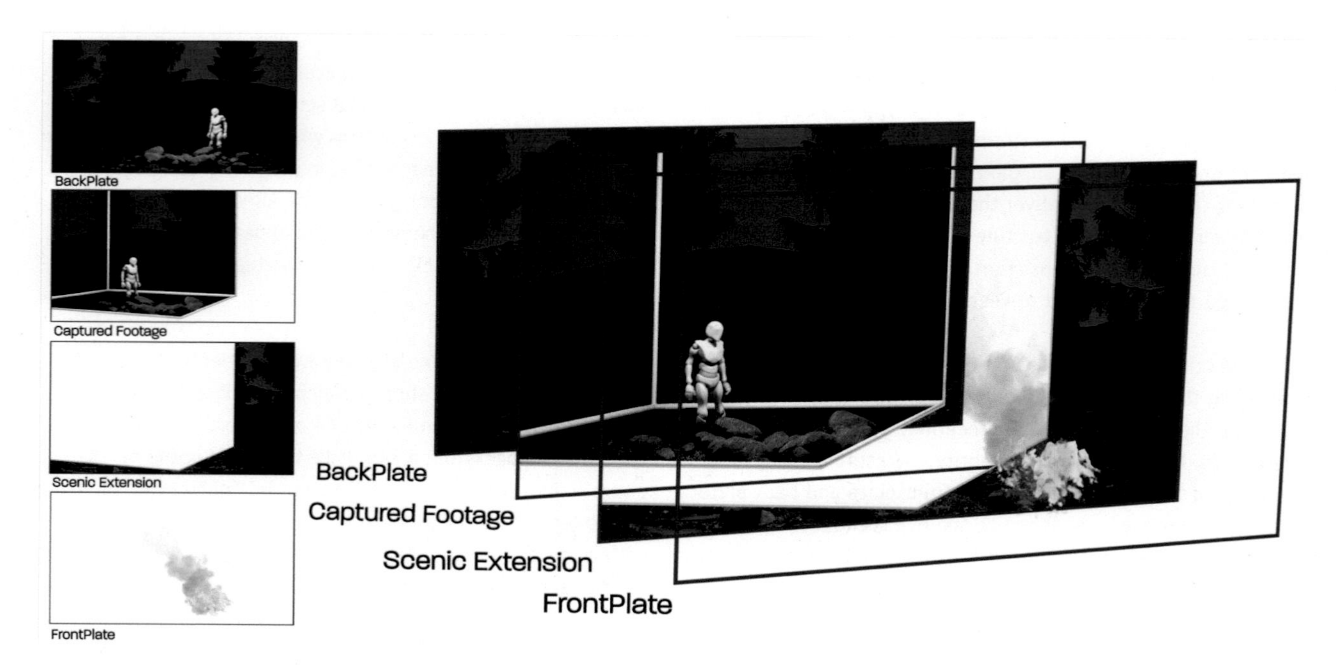

Figure 11.5 Layer stack with Scenic Extension Layer. Source: Image Credit: Ian Macintosh.

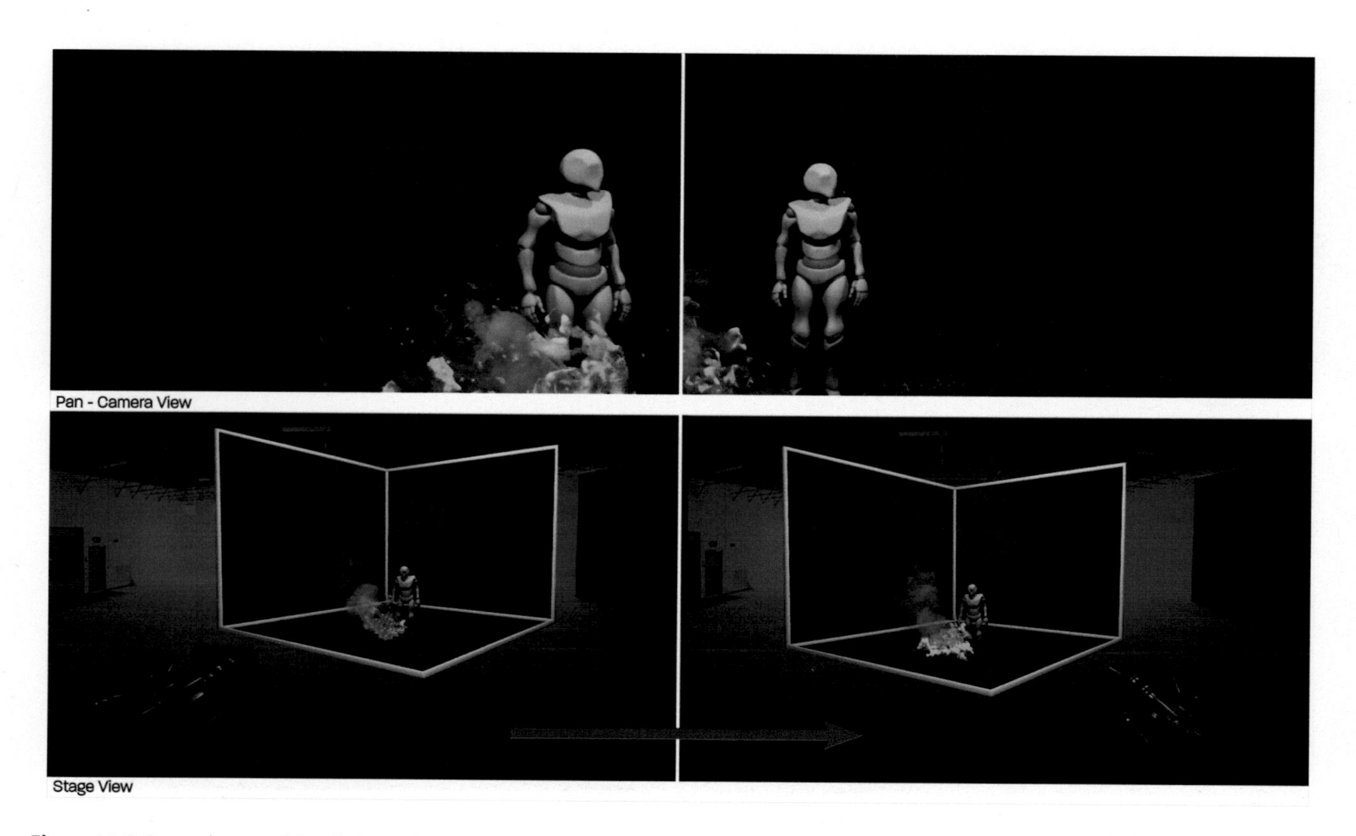

Figure 11.6 Pan and zoom of Scenic Extension must keep the AR layer of composite aligned with the LED stage. Source: Image Credit: Ian Macintosh.

Figure 11.7 Performer obscuring AR imagery based on performer position. Source: Image Credit: Ian Macintosh.

AR + Depth Sensing

What happens if our actor walks in front of the fire pit, obscuring imagery that exists in the foreground of our layering structure? This puts the actor in between the camera and the front plate content, forcing the front plate to appear in the background of the live action camera shot. Due to the actor's potential to vary their pace and position, we can only manage this illusion in real time (see Figure 11.7).

The stack of layers does not change. The AR fire is composited on top of the live captured footage. Instead of changing layers, we need to account for which pixels exist "behind" the performer and not render those onto the front plate (see Figure 11.8). Due to the interactivity of the physical world on stage impacting the content rendered in the virtual world, we would consider this an example of Mixed Reality (MR).

There are different technologies that can track the performer's location and define which pixels need to be eliminated from the front plate to make an object like our fire pit appear as though it is in the background. Depth sensing cameras like Intel's RealSense can locate the performer in the 3D world managed in the real-time AR rendering system. The real-time content platform would then interpret this information to define which pixels to leave undrawn in the front plate. MoCap and IR trackers can be used in the same way, or color could be used, similar to many sports AR processes. In time, I expect Machine Learning algorithms to deliver the best result for this type of real-time compositing.

Mixed Reality

What happens if our actor blows on the AR fire pit to get it started? Could we add a sensor in the fire pit to collect data about speed and direction to impact the digital fire imagery that is generated? What if the actor lit a torch in the fire pit and walked across the stage? An IR locator beacon could be added to the end of the torch and tracked. That location information would not only inform the front plate for additional AR fire on the torch, but also impact the back plate as the warmth of the fire in the torch affected the color of the background.

When the real world actions on stage impact the content generated in the virtual world, this is Mixed Reality. This narrow distinction separates MR from other VP tools, as

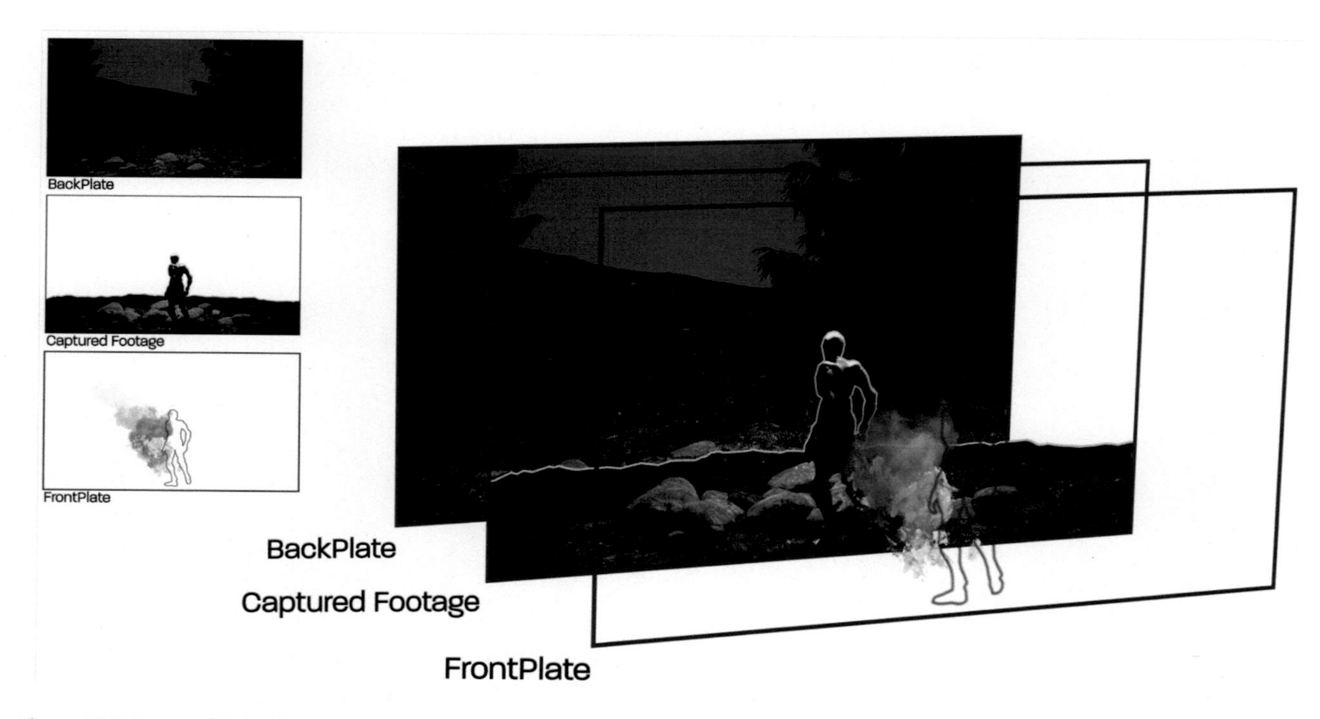

Figure 11.8 Layer stack showing what pixels must be eliminated from the front plate. Source: Image Credit: Ian Macintosh.

the interactivity is specific to actions taking place that are captured on camera. The practice of Virtual Production is generating interactive content, determined by camera movement. Therefore it is necessary to clarify MR as *on camera* interactive content.

Mixed Reality and Extended Reality are powerful and complex production tools. As I write this, Virtual Production teams continue to gain expertise and advance the practice of XR and MR at a dizzying pace. The software and hardware platforms that support real-time VP workflows respond to new approaches and industry demands as fast as they can. Studios are challenged to commit to gear, stage type, and expertise that is changing as fast as it can be acquired. Continuing our discussion from Chapter 10, let's consider some important reasons a production might choose one VP stage type approach over another.

LED vs Green Screen, Part 2

We've reviewed differences between LED stages and green screen for Virtual Production. Here, we examine the specific factors that will determine the decision to use one approach over the other.

These factors include:

- Environmental Immersion
- Lighting and Reflections
- Budget
- Speed
- Camera Frustums
- Background Cohesion

Environmental Immersion

An acting style for green screen performance has emerged over the preceding decades as actors have partnered with CGI characters in CGI environments that they may never see or experience. Creating natural responses to such invisible elements is not a skill every actor possesses.

On the other hand, when acting on an LED stage or projection surround, performers become immersed in an interactive live environment they can see and hear. This inspires more natural responses to events that might otherwise be hard to "feel."

For example, sci-fi films often put actors in situations that have no real world equivalent. How do you respond to a spacecraft making a light speed jump? If screens simulating the director's idea of a light speed jump surround a group of actors, you stand a much better chance of seeing a natural and coordinated response in their performance.

Lighting and Reflections

Another advantage of LED screens are the natural reflections created on people and props. The small LED panels are easy to locate around a studio when used as a reflection source. Green screen casts a green tone on surfaces. This is especially challenging to remove from glass, metallic surfaces, and skin on performers. Emissive surfaces, like LED screens, provide light, color, and texture matching the intended background.

Imagine a closeup of an actor driving down a curvy highway at night at high speed. While the production could shoot on location, it is much safer to shoot on a stage. In this example, the director wants to shoot the performer through the front windshield of the vehicle. On the glass, we need to show the reflection of the street lamps while lighting and dimming the actor's face to match. Well-placed sections of LED screen can do both while remaining outside the camera frame.

There are circumstances where LED screen provides needed background imagery, but illuminates faces or props at undesirable angles. Floor LED panels are challenging sources of facial lighting that is an unnatural and often unflattering angle. The desired camera shot will determine if LED flooring is needed.

Budget

An important factor to consider is budget. Cameras and production staff, gear and time, all have significant budget considerations. You can assemble a VP stage as expensively or as economically as possible within some boundaries.

Green screen is the most affordable approach. Anyone with a green backdrop, fixed camera, and a couple of good lighting sources can broadcast live from their living room and have it appear as if they are on a beach. In a studio setting, VP green screen can function well with a small team and limited production time. The technology to support VP green screen is more affordable and easier to calibrate and maintain during a shoot.

However, the image quality of VP green screen currently does not match the output of an LED stage when used for live broadcast. Without the emissive properties of LED, the people and objects on the green screen stage are lit differently than the virtual world around them. Depending on the scene, this may not matter, or we can, given time and budget, correct it with specialized lighting.

I expect we will see real-time image enhancement tools designed for use in green screen studios. Image processing of real-world objects to appear naturally within a virtual environment is an ideal AI task. Color correction and other image clean up processes are time consuming post-production jobs. I see machine learning algorithms providing real-time solutions in the not too distant future. This will put VP green screen workflows of high quality within reach of many more users with excellent results.

LED stages increase the budget with added gear and engineering time. LED wall rental, screen engineers, and camera calibration time all add to cost. Calibration may need to be repeated regularly through production. When this occurs during shooting, that's a lot of high dollar day rates waiting on the clock. Many productions also run on a tight schedule that cannot accommodate or afford lost hours to unexpected VP issues.

However, the results are likely worth the expense and not every production suffers time or budget crushing technical issues. For film productions with hefty post production budgets, some of those funds are more economically used during shooting using VP, and might save a production money overall. For live broadcast and events, LED stages replace sets and large lighting rigs, but producers must still account for virtual scenery and lighting in content creation. These budgets have historically been under funded and producers see VP as too expensive. Still, many live productions are using this technology and adapting to the costs.

Speed

How fast do you need to get this project up and running? How much production time is there onsite? While budget concerns often correlate with speed concerns, these are import questions to consider while looking at your budget for Virtual Production.

Ultimately, green screen is a simpler technological approach and a skilled team can get a new project up and running quickly. LED stages integrate more technology, requiring more setup time, testing, and calibration. On an existing LED stage that has an experienced team who has worked out any issues in the studio, you might find the working speed comparable to that of a green screen studio.

The important point to remember, as a new field, many teams are still gaining experience with the platforms and technology that make XR possible. Speed is not a guarantee working in any VP discipline. Real-time content creation tools and LED backgrounds do not provide speed because computers are doing more of the work. The computers are run by people who need experience, preparation time, and a well-planned Virtual Production schedule to realize any speed gains over a traditional production.

Background Cohesion

As noted in our earlier example of the forest fire pit, LED stages have edges. If that edge is visible in a camera shot, the content in the screen has to be stitched together with an AR scenic extension to complete the background. This technology depends on careful camera calibration and highly controlled camera settings to work.

Projects using a large LED Volume to maintain LED in the shot's background or those that don't require compositing live camera feeds for broadcast can avoid this process. Alternatively, green screen doesn't face this issue since objects that obscure the color green in a particular area define the background edge. The live compositing process can continuously fill the background even when the edge of the green screen is in view of the camera.

Can your budget afford more LED or more time? If you can keep the LED screen in your camera background, you will save time. Do you have access to a small box style LED stage at a good price? Spend that budget on calibration time and experienced XR engineers. Do you have more than one camera to track? Keep reading.

Camera Frustums

The camera frustum is the invisible boundary that defines the edge of the camera lens to the screen surface of the background. In computer graphics, the view frustum is analogous to the camera frustum (see Figure 11.9).

When using more than two VP cameras at the same time, we must manage the frustums so that they do not overlap when using an LED stage. With green screen, this does not present an issue as the generated imagery is not visible to the live camera. On an LED stage, the active camera determines the back plate in use on the LED screen, or else the frustums must not collide.

When multiple VP cameras shoot on an LED stage, operators can choreograph the cameras to prevent the frustums from overlapping. Alternatively, we can use complex switching to change the active camera exactly with the change of signal going to the LED stage. Sync and delay must be precisely managed so that the LED screen changes signal sources on

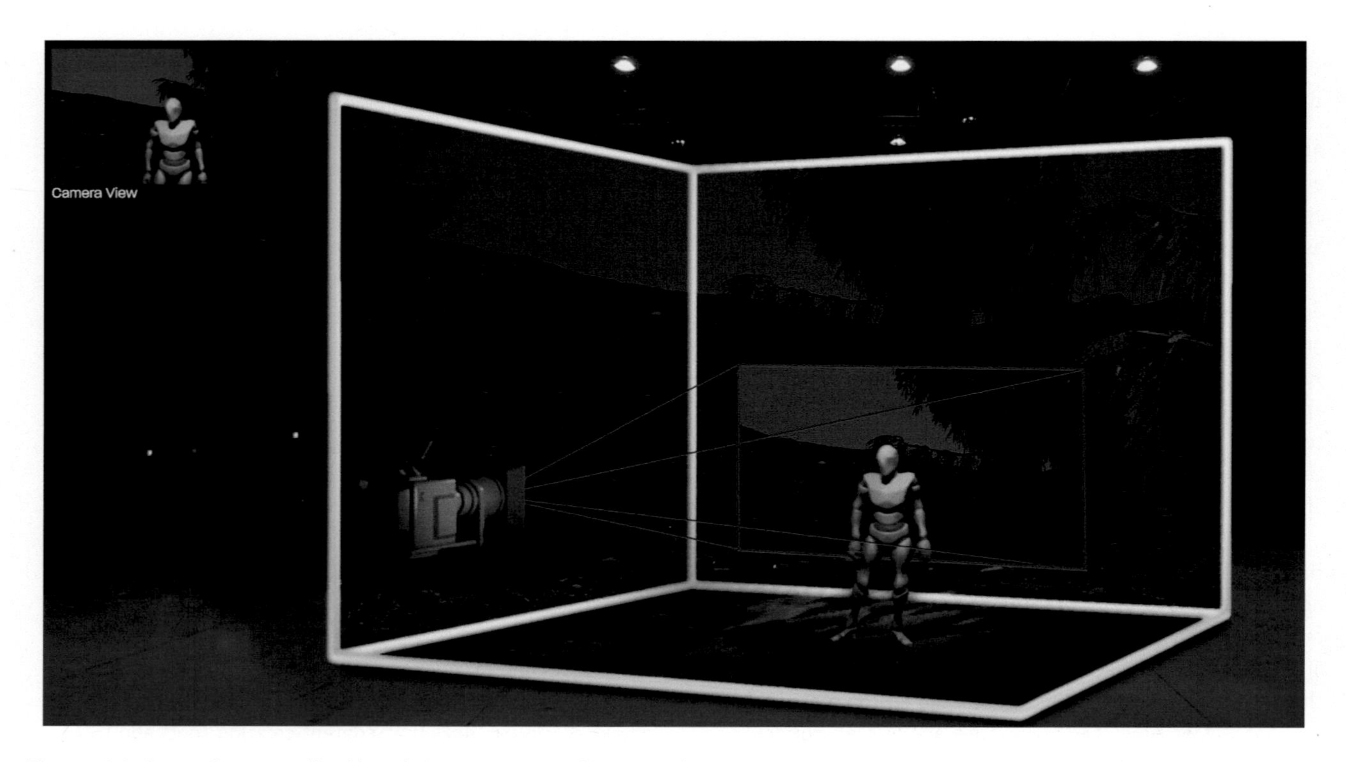

Figure 11.9 Camera frustum outlined in red. Source: Image Credit: Ian Macintosh.

the exact same frame that the active camera becomes live to broadcast.

There is new technology available that syncs the camera to a fraction of the available cycles of an LED wall. If a camera captures footage at 60fps, and the LED wall outputs at 240fps, sync offsets between cameras mean up to four cameras can see completely different backgrounds. Four video signals interweave to the screen simultaneously so that a frame from each signal plays in order.

Learn more about multi-camera frame sync technology at rtv-book.com/chapter11

Background Quality

Finally, there is image quality to consider. What is the demand on the production for final image quality? The best image quality uses a combination of both LED stage and green screen technology. The best parts of both tools combine to make a real-time ICVFX process that is finished in post-production.

In this process, an LED stage provides all the lighting and reflections, but the background captured in camera is green (see Figure 11.10). The camera path is recorded and delivered with the footage so that the VFX team can key a high quality CGI background into the background of the shot. Since the green is always behind the physical elements in the shot, there are limited green reflections to manage. You can also use a lower quality LED wall as it only serves as a lighting tool and a dynamically located green screen.

Any backgrounds created in real time, whether for green screen or an LED stage, are only as good as the artwork, modeling, and the rendering hardware. LED stages also depend on the quality of the LED wall, which are a function of resolution and LED quality. Quality of LED varies in color rendering and function at low brightness, which becomes visible in shallow gradients. LEDs suffer from a poor ability to provide a range of low levels. This is exacerbated by signal compression that also limits range for shallow gradients.

Models should be optimized for the rendering hardware in use to maintain necessary frame rates. This goes hand in hand with communicating potential real-time hardware limitations to the content artists so they can also optimize the project files for efficient rendering.

Even with all these considerations, these platforms are within reach for a wide range of budgets. As producers adapt to Virtual Production, money allocated for post-production might be

Figure 11.10 Green screen workflow combined with LED stage workflow. Source: Image Credit: Ian Macintosh.

better spent on VP production. For live broadcasts, budgets must adapt to take advantage of VP tools that might offset existing production costs, though most productions currently find VP to be an additional expense. When a project is planned for VP in its entirety, the results can be quite exciting, as with this project from Lux Machina.

Case Study – Lux Machina – Worlds 2020

The League of Legends World Championship, also known as Worlds (Figure 11.11), is an annual esports event produced by Riot Games, the maker of the League of Legends video game. From September 25, 2020 to October 31, 2020 the 10th Worlds was held in Shanghai, China in the middle of a global pandemic. Typically, Riot Games hosts a multi-city tournament that culminates in the finals, a stadium-level event often with 50,000+ fans in attendance and tens of millions watching worldwide, to represent the global nature of our sport. Because of the restrictions caused by Covid-19, Riot tasked creative producers Possible, Inc. with coming up with a way to present Worlds in a virtual environment while maintaining the dynamic nature and grandiose delivery that the tournament had become known for.

Possible and the Global esports Team at Riot conceived of four different worlds inspired by the skyline of the host city and the four elements that had been introduced as a game mechanic for the pro season. These worlds would house each stage of the tournament to echo the touring format of past exhibitions. Possible designed the worlds from the ground up, building living, breathing environments where the camera could look in any direction. 3D assets were modeled in Maya, FX were simulated in Houdini, and texturing, lighting, and scene assembly were completed in Unreal Engine.

Lux Machina Consulting (LuxMC), a company known for their work designing the LED Volume and Virtual Production workflow used in Season 1 of *The Mandalorian*, was brought on as technical producers to deliver Possible and Riot's creative vision. LuxMC started by designing the stage that the event would be held in, a cube-shaped volume with LED on three of the walls and an LED floor (see Figure 11.12), totaling over 900 tiles of LED displaying real-time rendered visuals at 32K resolution and 59.94 frames per second. The volume was supported with 20 custom-built render servers, purchased by Possible, running Unreal Engine 4 with heavy customization from the LuxMC programming team. Four RedSpy Stype camera trackers were used to track two cameras that offered dual frustum capabilities to create a realistic parallax effect behind whatever they were shooting.

The first challenge that the team had to address was allowing multi-frustum camera support and integrating its use into the production. The Director and AD, who were operating from a remote location, were given a control that would send an OSC command to an IO Core. Custom blueprints were created in UE4 to receive this command which, in turn, would change which of the two cameras were projecting their frustum onto the LED wall (see Figure 11.13). This signal had to be delayed and synced to the exact frame of the actual camera switcher so that the audience would never see a frustum changing from one camera to another. Additionally, the camera operators, who had not worked on a Virtual Production stage before, had to be trained on setting up a shot without being able to see their

Figure 11.11 Camera shot from Worlds. Source: Image Credit: Riot Games.

Figure 11.12 Three walled LED stage of Worlds, during a promotion shoot. Source: Image Credit: Riot Games.

Figure 11.13 Working with two active camera frustums on a digital stage. Source: Image Credit: Jake Alexander, LuxMC.

projected frustum or scene extension before their camera was switched to.

Another technical obstacle was creating the system that supported interactivity between the game and the volume. OSC triggers were sent from the game to the servers which would produce an effect in real time in the AR environment or on the LED wall. For instance, when a player would select one of 140 different characters on the character selection screen, an animation would trigger on the wall behind the players. There was also a virtual scoreboard that floated in the set extension above the stage. Hundreds, if not thousands, of triggers had to be programmed to custom blueprints in UE4 to achieve this effect of player interactivity.

The final hurdle that Possible and LuxMC needed to overcome was how to produce 240 hours of live-live content with 0 on-air crashes. The software and hardware that was being used was still fairly untested in a broadcast environment and this would be the most complex version of deployment to date. LuxMC and Possible worked with the production team to create a contingency plan in the case that the system went down in the middle of broadcast. The plan was to cut to a still store on the E2 or directly to in game coverage if possible. The technical team would reset the system and the camera crew would make sure not to shoot any of the set extension or other AR components until the servers were restored. This required the entire production team to be in lock step with the technical team to create a seamless and undetectable transition from a compromised system back to a functioning one (see Figure 11.14 for behind the scenes activity).

Due to time and logistical restrictions, testing for this system was very limited. Technical tests began roughly a month before the start of the event, where a section of the final stage was built in Los Angeles so the pipeline could be vetted at scale. The schedule was meticulously plotted out so that the render servers were delivered in batches straight from the manufacturer, allowing testing to ramp up as the servers became ready. This time was mostly used to confirm that the outputs on the servers were functioning properly and target frame rates in the Unreal scenes were in striking distance. The

Figure 11.14 Behind the scenes running the Worlds stage. Source: Image Credit: Riot Games.

Figure 11.15 Jib operator, Dawn Henry, controlling the camera head on the jib. Source: Image Credit: Riot Games.

day after testing wrapped, the team departed for Shanghai in order to begin a mandatory 14-day quarantine, two weeks that were spent programming and testing the OSC triggers that would be used for player interactivity while Unreal scenes were further optimized to make broadcast frame rates viable.

As with any project of such complexity, workflows needed to be altered in order to marry the creative vision with the capabilities of the hardware. One instance of this was the pre-taped Augmented Reality performances that preceded the quarter finals. When the team realized that the massive AR load of one of the scenes could possibly affect the performance of the render nodes, a plan was put into place to record all four outputs of the pre-composited feed in case it was necessary to composite those feeds in post rather than during the recording. This change did not affect the final project in any way, only the path to get to it.

The end result of the efforts of Riot Games, Possible, and LuxMC was an event that was unlike anything that had come before it. The team produced 240 hours of live content with 0 on-air crashes. There were a total of 12,500 camera cuts, 32,000 multi-user transactions, 7.8 million OSC messages sent and received and a total of 52 million frames rendered at 59.94. Worlds 2020 was the most advanced Broadcast Virtual Production setup to date and has paved the way for future live-live Virtual Production events (Figure 11.15).

by Jake Alexander for Lux Machina Consulting
www.luxmc.com/

LuxMC develops and engineers leading-edge technical video solutions for film and TV, broadcast, live events and permanent installations. We specialize in Virtual Production, in-camera visual effects, display technologies, and creative screens control.

Project: The League of Legends World Championship
Location: Shanghai, China
Year: 2020

Project Credits:
Executive Producer: Wyatt Bartel
Producer: Sheiva Khalily

Client:
Possible Productions for Riot Games

CHAPTER 12

The Future of Real-Time Content

We are in a transformational development cycle of the many facets of real-time content creation. Entertainment projects of all varieties find and apply new technologies to creative video design and Virtual Production, advancing the real-time content production process. There are many opportunities in this field that are undefined as of the writing of this textbook. While we can't predict the future of this corner of the entertainment industry, what we can do is outline skills that will be valuable for any of the roles that are expected to develop in this field.

We have discussed real-time video content use in the film industry, video art, television broadcast, interactive installations, live entertainment, and more. The skills that drive real-time technologies that serve these production environments vary significantly: designers, 3D modelers, 2D content creators, coders, producers, and so on. We'll look at a list of commonly demanded skills at the end of this chapter.

We have also seen several case studies of creative video production teams applying these skills. This represents work from designers, creative technologists, engineers, producers, content artists, and more. Many of these examples involve technologies known to these professionals, but applied in new ways, often for the first time. Occasionally, new skills are learned in the process of building a project. The demands on these professionals require thoughtful planning, testing, and occasionally, failing. Failure is a part of this work. What makes a team successful is finding solutions after a plan doesn't go as planned.

As you review these case studies, consider the challenge presented and look at the ways these professionals go about researching and developing solutions. Each is unique. There is no set pathway to success when using real-time content creation. This is an evolving practice that requires constant learning. Each job is its own research project.

This textbook will point readers to important software and systems to learn. No one can be an expert in all of them, but awareness of how they work will inform planning choices. Onboarding a project, you might see that a design requirement is best suited for a tool you don't know well, rather than the tool you do know. That's a good thing. Better to hire a team member with the right skill or hand off a project to a more experienced team member than to not deliver the project. Focus on being excellent at the skills you know and have a general understanding of the capabilities of other solutions.

We as practitioners of real-time content creation have to serve the entertainment technology landscape that exists now and plan for the skills demand that might come next. Understand the skills you have and where you want to develop them to be useful for future practices. We have discussed emerging technologies and industry trends and we will review those and a few more. You can find the most current and groundbreaking work in this textbook's companion website.

> Visit rtv-book.com/chapter12 for resources to find the latest developments in real-time content creation. This will be an updated resource you can refer to as needed.

Developing Industries

We have devoted much of our recent discussion to Virtual Production. These are some of the most complex applications of real-time content creation and require careful study. However, we've also seen ways to leverage real-time content creation for any creative video project. A project that uses creative coding or traditional motion graphic production can also make use of real-time content creation tools as we will explore in detail in this chapter's case study.

We've discussed the convergence of the internet, gaming, and entertainment production. The tools that drive these individual sectors cross over to the other to advance production

DOI: 10.4324/9781003206491-14

capabilities in new ways. One specialized entertainment industry application where we see this crossover is online performance events.

Many attempts have been made to use the internet as an interactive performance landscape, but only a few stood out as something beyond television broadcast to the web. Marshmello performed one of the first online in-game concert events in early 2019 on Fortnite. However, the breakthrough performance that merged the game and concert experience occurred in April 2020 with Travis Scott. Other performances soon followed, expanding to other game platforms, with live audiences consistently in the tens of millions.

See video of these events online at rtv-book.com/chapter12

Productions also repurposed game engines to hold online music festivals. Tomorrowland EDM festival was rebuilt from the digital ground up to host its 2020 festival as a virtual event. Attendees could visit different stages and navigate the virtual crowd to experience their favorite DJs perform. Professional conferences went online as well. Interactive web communication tools gave attendees the power to pose questions to speakers and gather in chat spaces to connect. Virtual events continue to expand online for job fairs, education seminars, trade shows, and virtual malls. As more entertainment production companies pivot to understand their place in the metaverse, virtual event experiences will dominate as they continue to mature.

The return to in-person audience live events has not slowed down the pace of online event development. There is an expectation that attendees will pay for virtual access to music and conference events going into the future. Improving the audience experience is a top priority, so that virtual attendees feel more present in a hybrid event format. Audiences want more than participation via a video conference window. They want to be there, remotely.

While the entertainment industry works to solve hybrid audience entertainment, I expect to see more previsualization via a web browser. The use of 3D on the web is more than a decade old, but adoption has been slow. In the last few years, there have been more commercial websites using 3D in their product demos or real estate viewers, and the user base for web 3D is advancing beyond gamers. Interacting with models on the web is more common.

I see 3D on the web as the next stage of online production communication. 3D visualizations are information rich and interactive. The viewer has full control of their point of view of the information that is shared, right in the palm of their hand. Entertainment production teams distributed around the planet are dependent on remote communication. Flattened images of studios and stages are as difficult to decipher as a book about video. We need to communicate in meaningful ways with simple and highly informative resources. I believe more companies that support VP platforms will support web-based sharing tools.

Between online event production and 3D based communication and visualization, more of the real-time content tools we use will become the technology backbone for the metaverse. What we mean by "screen" might have more to do with the glasses you are wearing than the surface you are watching. This continuing advance to a new entertainment future, whatever it looks like, will have event producers looking to us for solutions.

Developing Technologies

The most important real-time content creation tools for entertainment applications to learn today are Notch, Unity, and Unreal. In the fields of computer graphics, computer vision, image rendering and image generation, new technologies will always play a role in how creators build content in real time. Advances in machine learning, neural networks, and AI to generate imagery will impact entertainment video content creation. I recommend attending a SIGGRAPH conference or viewing their catalog of presentations at their website to see the variety of work being done in an assortment of image generation technologies.

How do you know where to invest your time in learning? I recommend that you find something that engages you about the process of creative video production and start learning to code or create imagery with that tool. There is rarely a time that learning a skill for content creation has become un-valuable. Adobe Flash is a good example of an animation tool that is no longer used, but the people who built projects in Flash have learned essential problem solving and project building skills that apply to any tool. Software and hardware go in and out of fashion, but understanding how to plan and execute a project applies to any tool.

Broad knowledge of what tools are available, their capabilities and costs, strengths and limitations, is always valuable. I repeat: no one person can have expertise in all the tools. However, some insight into the larger world of content creation tools is essential for good project planning. Don't shape a project to meet your current skills. Be prepared to learn or hire team members with the right expertise for a project.

When a client calls with a request that sounds like, "What's the latest technology?" I consider this a red flag. Projects should start with good creative. If the desired creative goal has yet to be achieved, that is an opportunity for innovation or compromise. Technology develops in response to new vision, as long as time and budget can support that vision.

If you are curious to know the latest technologies, go to the source and see what research is being done in computer graphics processing. Explore the methods used by video artists that inspire you. Video card manufacturers to universities, VFX companies to creative agencies, are all doing research and development for the future of image generation. I have posted resources on the chapter companion web page. In the meantime, consider these hard skills:

Real-Time Content Creation:
Unity, Unreal, Notch

Content Creation:
After Effects

VFX Creation:
Nuke, Maya, Houdini

3D Modeling:
Cinema4D, Blender and 3DS Max

3D Drafting:
Vectorworks, SketchUp and AutoCAD

Creative Coding:
Processing, C#

Traditional Coding:
C++, Java, and Python

Programming Media Servers and Real-Time Content Platforms:
disguise, TouchDesigner, Pixotope, Pixera, Smode, Green Hippo, ioversal, Modulo Pi, Real Motion, vizrt, VYV Photon, Zero Density and others

Technical Production and Programming Skills:
GrandMA, Artnet, Networking, Camera Tracking

Designer Skills:
Lighting, Scenic and Graphic Design, Animation Design and more

Overwhelmed? Don't be. Pick up something and start. Everyone was a beginner at some point. Everyone has struggled to learn these complex tools and apply them. Stick with it, keep learning, find opportunities to practice, and you will build projects like the World Expo in the following case study.

Case Study – Bild Studios – Al Wasl – The Interactive Experience

How Bild turned Al Wasl Plaza into the world's largest interactive installation for Expo City Dubai

Summary

London-based media technology company Bild Studios were commissioned to turn the world's most complex 360 projection installation, the Al Wasl Dome in Dubai, into the world's largest interactive audiovisual experience. Running real-time content across a 27,000 x 6,000 pixels canvas, featuring 252 projectors, 16 disguise media servers, 27 line arrays, all controlled from an iPad, and working closely with the creative team of Al Wasl Dome, the team at Bild used their extensive knowledge in real-time workflows, video engineering, and creative design to realize the huge task. As a result, the work would not only create a jaw-dropping immersive experience (see Figure 12.1), but also reduce the need for producing new and unique content that takes up a vast amount of hard disk space.

Background

World Expo 2020 presented an important opportunity for Dubai to excel on the global stage. Celebrating innovation and industry, the event is an opportunity for all nations to connect and share technical and commercial progress, showcasing world firsts in architecture, technology, and the arts.

The centerpiece of the Dubai Expo was the iconic dome structure of Al Wasl Dome – a visual marvel represented as a 130m diameter by 70m tall dome structure (see Figure 12.2) which hosted the opening and closing ceremonies and 173 special events and spectacular functions throughout the six-month celebration of Expo.

With a total of 252 Christie 40k lumen projectors covering a surface over 25,000m^2 across a 27,000 x 6,000 pixel canvas, the

Figure 12.1 Live-actions shots from Al Wasl – The Interactive Experience. Source: Image Credit: Bild Studios.

Al Wasl dome represents the world's largest and most complex projection surface to date (see Figure 12.3).

Bild was integral in the installation of the Al Was Dome and was commissioned back in 2019 by Creative Technology Mid-East and later directly for Dubai Expo 2020 to produce complex feasibility studies, technical analytics, and the creative content pipelines to bring the huge dome structure to life, using the previously mentioned 252 projectors, with video signals fed from 84 unique HD channels. Anyone creating content for the dome today is using the workflows Bild produced. An updated content creation pipeline was later created which allowed artists to work with 3D, 2D, and real-time content in parallel and allowed them to merge them into an even more seamlessly integrated content pipeline than the original version.

Working directly for Expo, Bild continued to work on a number of high-profile projects in the dome throughout the coming years. Examples include creating the Augmented Reality (AR) broadcast overlays for six tracked cameras for the opening and closing ceremonies, developing a photo-real VR visualizer, integrating Unreal Engine into the projection system, as well as creating bespoke projection content for selected events (see Figures 12.4 and 12.5).

Figure 12.2 The iconic Al Wasl Dome is the centerpiece venue of the Dubai Expo 2020. Source: Image Credit: Expo 2020.

Al Wasl – The Interactive Experience

Shortly after completing the original content pipelines, Bild were engaged by Expo to explore another pioneering opportunity – to transform the space into the world's largest interactive real-time installation.

The creative team at Expo, headed up by Expo City Dubai Executive Creative Director Amna Abulhoul, had a vision with two main objectives in mind; to first and foremost create an impressive and highly immersive VIP experience where chosen individuals could stand in Al Wasl and generate stunning real-time visuals within the impressive architectural structure. The second objective was driven by much more of a practical nature; to lean on real-time generated content, created in the engine Notch, to save time on pre-rendered content ingest and reserve hard drive space within the busy disguise media server system. In fact, the Expo team realized soon that many real-time based looks could be embedded in the daily shows, and utilize the freedom of adjusting content in real time rather than testing out pre-rendered content across multiple time-consuming iterations. Doing so removed the need to create new and unique pre-rendered content, for example for background loops and ambient graphics in-between the main show sequences (see Figure 12.6).

In addition to the visual design aspect there was also the opportunity to generate interactive audio for the in-house L-Acoustics audio system, made up of 27 discrete Loudspeaker arrays and six subwoofers distributed around the dome. Consequently, the final result would not only represent the world's largest interactive visual installation but also using one of the world's biggest audio installations – reacting to the finger movement of the app-user.

After being presented with a brief illustrating a series of treatments Expo wanted to create, Bild devised a four-phase delivery to bring the installation to life. Due to the extreme size and advanced specification of the installation, the realization of the project was the close relationship between

Figure 12.3 Outside views of the Al Wasl Dome. Source: Image Credit: Expo 2020 (top) and Bild Studios (bottom).

the advanced technical engineering and creative development – a process that had to be completed in unison.

Phase 1 – Technical Feasibility and Creative Exploration
Phase 2 – Creative Development and Audio Design
Phase 3 – App Development
Phase 4 – Onsite Installation

Al Wasl Dome Fun Facts:

- 130m diameter wide, 70m tall dome structure
- 25,000 sqm projection surface
- 252x Christie 40k lumen projectors
- 27,000 x 6,000 pixel resolution

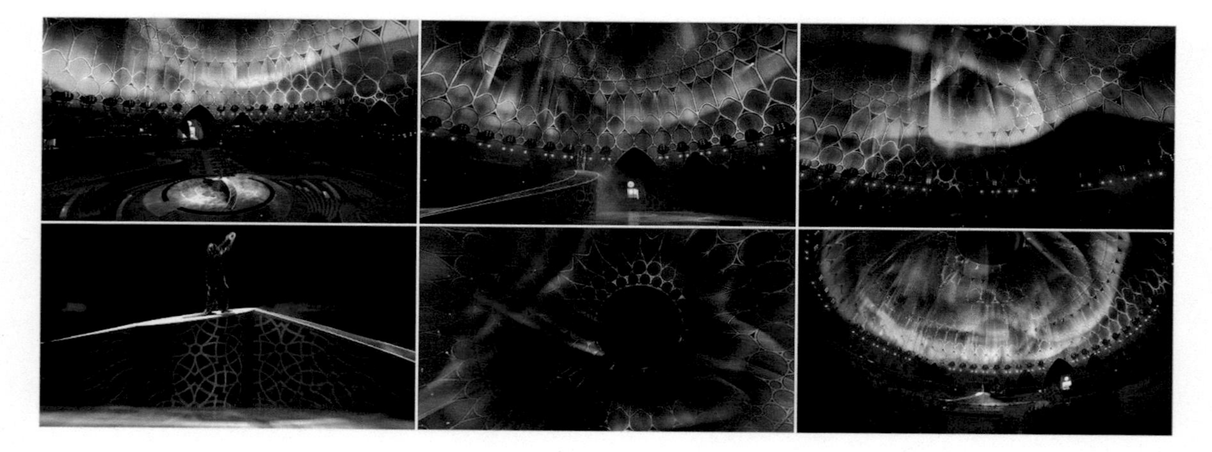

Figure 12.4 Images of the broadcast AR overlays for the opening ceremony of the Dubai Expo, created by Bild. Source: Image Credit: Bild Studios.

Figure 12.5 Images of the broadcast AR overlays for the closing ceremony of the Dubai Expo, created by Bild. Source: Image Credit: Bild Studios.

Figure 12.6 Example of real-time generated content from Notch – embedded into the daily shows to save hard drive space and with the added ability to adjust the content on the fly. Source: Image Credit: Bild Studios.

- Real-time-generated visuals using real-time engine Notch
- Powered by 16 disguise gx2 media servers
- iPad with custom-designed user interface

Phase 1 – Technical Feasibility and Creative Exploration

Background

Creating the fully functional and efficient workflows for the *pre-rendered* content was itself a great challenge and achievement by the Bild team. To kick off the process the Bild team carefully processed the projection setup, including the analytics of brightness levels, pixel sizes, and blend zones

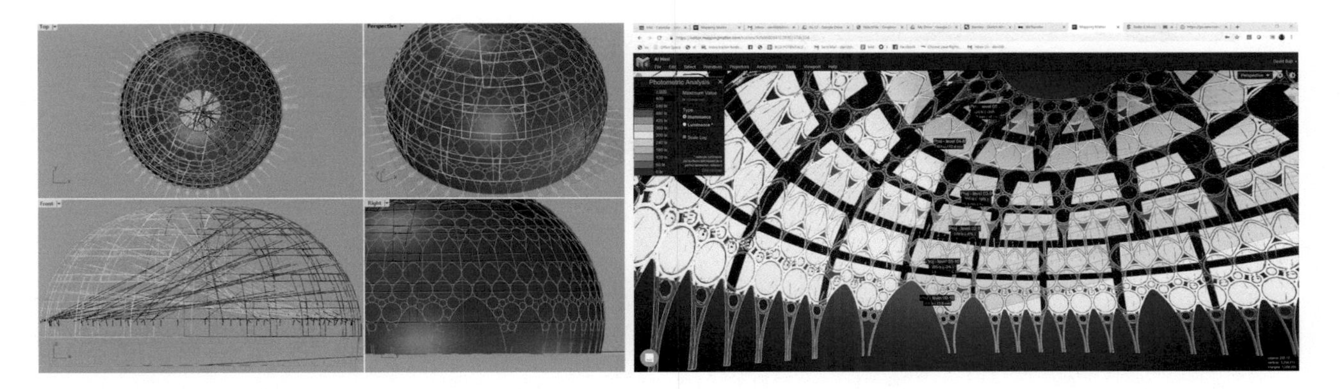

Figure 12.7 Images left and right show the initial projection studies. Source: Image Credit: Bild Studios.

Figure 12.8 These images show how 2D content distorts as it travels up through the dome. In addition, 2D content implies a clear start and end to a canvas which isn't desirable for 360 immersive content. Source: Image Credit: Bild Studios.

(see Figure 12.7). With this information, the team could get an understanding of the total resolution of the canvas and start the journey of producing the content pipelines.

Producing content for any dome always requires extra consideration – simply because the dome shape cannot be unwrapped into a perfect 2D template, the so-called *UV Map*. In other words, distortion will always be introduced when working purely with 2D graphics directly on a UV template (see Figure 12.8). In addition, a 2D workflow always implies a start and end to a canvas, when in reality content actually needs to flow seamlessly across the dome.

Hence the need to introduce a *3D workflow* that lets artists render graphics in a 3D software. This solution not only opens up huge amounts of creative possibilities but would also remove distortion and enable a seamless start and end to the 360-shaped canvas.

To allow both 2D and 3D artists to work in parallel, the Bild team produced a workflow that lets artists create graphics in a 3D software that also integrates into a 2D pipeline. That pipeline then allows the artist to convert the visuals from one UV format to another. For instance, a spherical lat-long format could be converted to an fisheye format, then to a fully compensated non-distorted UV layout of the sphere, to a third UV format – and vice versa.

The system then auto-generates the 11 video files, split up across the canvas, with the correct resolutions and naming conventions (see Figure 12.9). They also use resolutions that allow proxy files to be generated that are always divisible by 4, as most video codecs still prefer resolutions divisible by 4.

As such, 2D artists and 3D artists can work in parallel and composite their files together – knowing that the final video files will always be generated correctly.

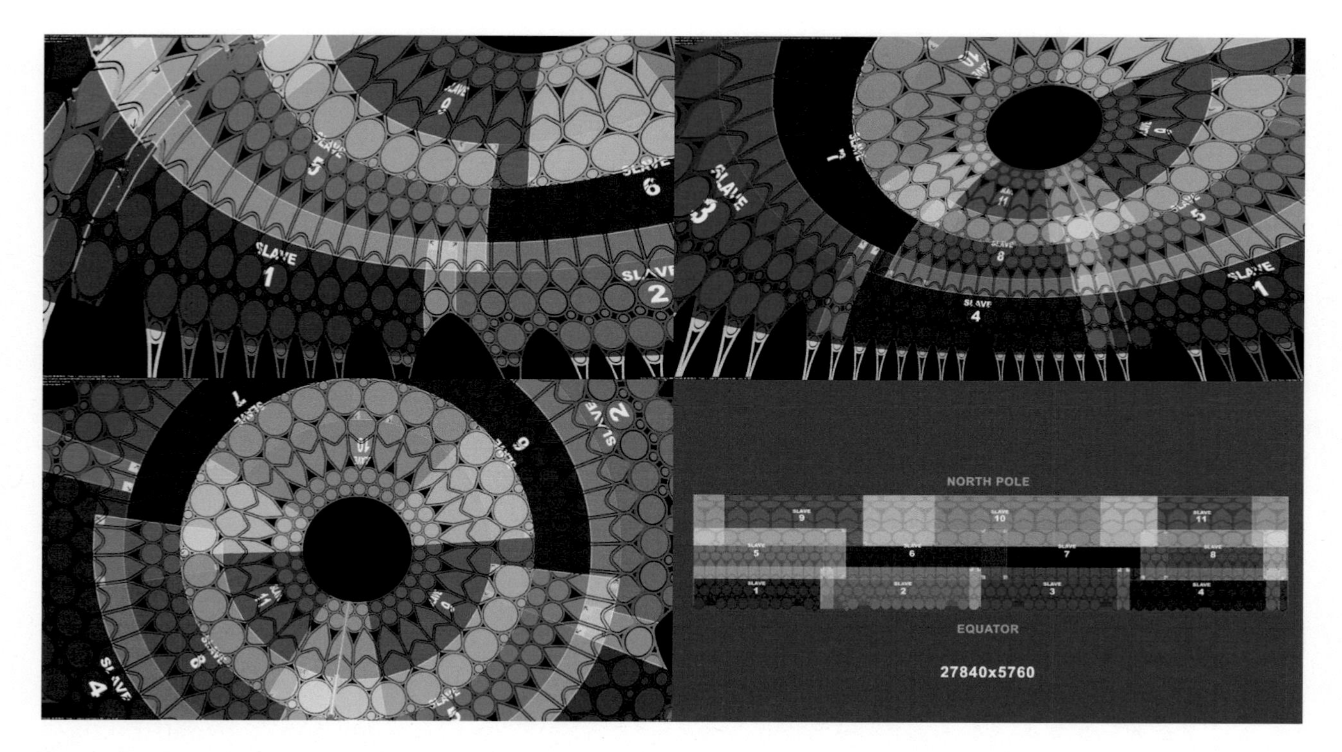

Figure 12.9 Bottom right: The 11 projection zones laid out on the UV map. Remaining images: Visualizing the projection zones on the dome. Source: Image Credit: Bild Studios.

Other considerations are rendering times and hard drive space. Depending on the content, a few seconds of content may take hours to render, and the final files would take up a huge amount of hard disk space on the servers, despite that only the relevant files would be copied to the corresponding servers.

The Need for a Technical and Creative Feasibility Study

Having the previous section in mind, converting the same enormous setup into a *real-time* driven solution, controlled from an iPad, may sound like a daunting task.

To overcome the challenge, the team considered the first phase of the project as a technical and creative feasibility study. Consequently, this phase centered around designing the real-time graphics pipeline as well as creating the first set of real-time visuals with focus on technical performance.

In addition, at an early stage, the team started to design and plan for the implementation of the iPad interface that the user would interact with in order to generate the visuals on the dome.

Choice of Real-Time Engine

With a 16 strong disguise gx2 server system already installed onsite, originally intended to only run pre-rendered content, the choice of picking a real-time engine that could bolt onto the existing setup wasn't a hard decision. The obvious candidate was the generative content creation platform Notch – purely due to its long and well-established integration with disguise.

For any reader who may not be familiar with the Notch-disguise workflows, it works as follows:

A Notch artist designs a real-time scene in the Notch 3D visualizer from which a Notch block is exported and imported into the disguise server. Disguise then renders the Notch block and produces the real-time visuals contained in that Notch block. If interactive input data such as slider values or finger tracking coordinates needs to affect the visuals (crucial for this project) the data would be streamed into the disguise server which updates the corresponding exposed properties in the Notch block. Disguise then renders the Notch block accordingly, usually 50 or 60 times per second.

The only other real contender for creating real-time visuals in this space would have been Unreal Engine. However, at the time of commissioning, at the early months of 2020, the Unreal-disguise workflow had not yet been developed and it would also have required a vast amount of additional hardware.

Requirements

It was imperative to create a Notch-disguise real-time pipeline that not only would allow artists to easily create stunning content but also make sure that the content would integrate seamlessly with the existing disguise mapping workflows that the Bild team had previously installed and verified onsite. In addition, with no opportunity for any form of pre-testing on the physical dome, mainly due to a tight Expo schedule and general Covid-19 travel restrictions, it was also essential that the performance and the sync between the servers could be fully emulated and verified in an office environment during pre-production.

Furthermore, to support fully interactive usage from an external device, in this case an iPad, the pipeline had to be able to support user input, such as tracked finger coordinates, slider values, and being able to push and pull data to and from the disguise project, that renders the Notch block locally on each server. For instance, if a user wishes to add more textures to the disguise system, they should automatically appear as thumbnails on the iPad interface without the need for any manual copying.

Splitting Up the Real-Time Rendering Power

In terms of available rendering power for *real-time* content it would certainly be impossible to render a 27k x 6k canvas on a single server. Also, with the 84 channels required to feed the 252 triple-stacked projectors at least 11 Actor servers would be required, in addition to the Director and Understudy (backup) servers. Therefore, the team asked themselves whether there was an opportunity to use the existing pipelines, based on splitting the canvas into 11 pieces, but changing playback of video files to real-time rendering of Notch graphics. Would the servers be able to perform, and would all servers stay in sync?

With 11 Actors in place, split up across a 27k x 6k canvas, each server would therefore "only" have to render approximately 7k x 2k resolutions. While this is still a big resolution it is dramatically less than rendering the full canvas.

Fortunately, after some initial testing (see Figure 12.10), and adding some bespoke functionality from disguise, the team proved that this worked very well. In fact, the team could truly emulate the onsite conditions in terms of sync and performance of the servers using a limited number of gx2 servers installed at the Bild London office. In more detail, three disguise gx2s were utilized in the office, emulating one Director server and two Actors. In terms of output configuration and software setup they were set up exactly according to the onsite environment. The Actors were always configured to be set up as adjacent to each other in terms of projection zones. Therefore, by running the same Notch scene on both Actor servers the blend between two servers could be monitored by capturing a selected HD output from each Actor server into a fourth disguise server (of any disguise model) that simply re-projected the visuals back onto the dome to its correct place. The blend between these two outputs was then continuously monitored. Should any content not sync or perform well, it was immediately recognized in the office rather than discovered onsite.

Previz

Key to the delivery of any creative work for the team at Bild, it was essential to pre-visualize the creative progress while

Figure 12.10 Photos from the first phase of the project which in large included a technical feasibility study verifying server performance and sync. Source: Image Credit: Bild Studios.

simultaneously proving and verifying that the treatment could be mapped and be operational in the real world. The team essentially needed to deliver a pipeline that allowed for client feedback on the visuals.

Dealing with the world's largest dome surface, which is partly transparent and with a huge amount of additional showlights, the team understood that it may be difficult to create a previz that would perfectly match the colors and brightness levels of the real site. Also, due to its huge size, it's sometimes hard to set the right *speed* of visuals – as viewing the speed of graphics on a computer screen often proves to be very different than in real life. Having this in mind it was imperative to inform the client that onsite color correction and tweaking of visuals certainly will be required.

With the impossible task of rendering the total resolution from a single gx2 server, 3 different visualization methods were deployed:

1. **View the content** directly **inside Notch.**
   This is the view for the artist when crafting the visual look inside Notch. However, while it would generate the best-looking quality visualization it wouldn't necessarily prove that visuals would perform and sync well. Therefore, this visualization method was heavily utilized during the early conceptual stages of the project (see Figure 12.11).
2. View the content across the whole dome at low-resolution, inside the disguise visualizer and full-resolution on an Actor server. As discussed, to verify sync and performance it was imperative to load the Notch block onto a gx2 server that would be configured to truly emulate onsite conditions. While this method of visualization didn't generate a high-resolution visual fidelity, it did prove that the visuals would perform well as well as receiving interactive data correctly from the external iPad.
3. View the one project zone (out of 11) of content at full-resolution in an Actor server. As a third option, the team could choose to view a dedicated projection zone from any Actor server. This would generate a 1:1, full-resolution view, while simultaneously proving that the visuals would perform well (see Figure 12.12).

With Covid-19 in full blast, with all team members working from home, the team demonstrated the latest progress of the client mainly via a live screen share in video calls. This was done by controlling the servers remotely using the app combined with the previously mentioned previz capabilities. By showing the visuals rendered on the fly the client could comment and update the treatments live during the call (see Figure 12.13). It shows how effective and powerful real-time content is compared to the traditional pipeline where clients usually have to wait days to see new updates.

Phase 2 – The App

With such high-profile guests visiting the Al Wasl Dome it was imperative to design a slick and simple app that they easily could interact with. Having that in mind, the team had to avoid using interfaces such as OSC or MaxMSP that come with a library of existing buttons, sliders, nobs, and other predefined UX elements. These interfaces tend to look very technical and would be limited in terms of functionality.

Instead the team eventually settled on creating a bespoke web design interface on an iPad, using standard HTML and CSS. This approach resulted in a simple and user-friendly UX design with

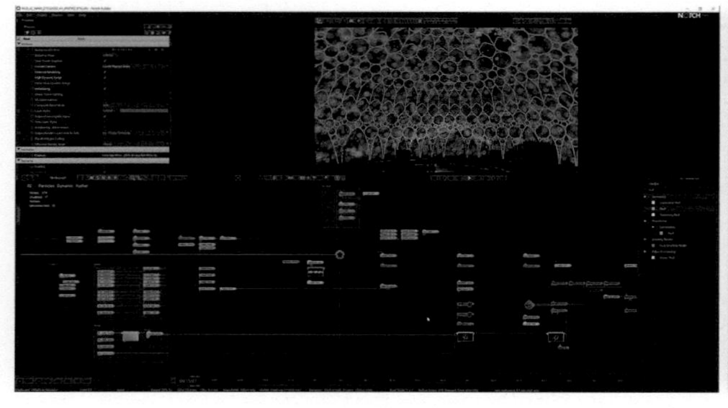

Figure 12.11 Real-time Notch visualization. Left image side shows the user interface of Notch with real-time visuals rendering. Right image shows a zoomed in real-time visualization inside Notch. Source: Image Credit: Bild Studios.

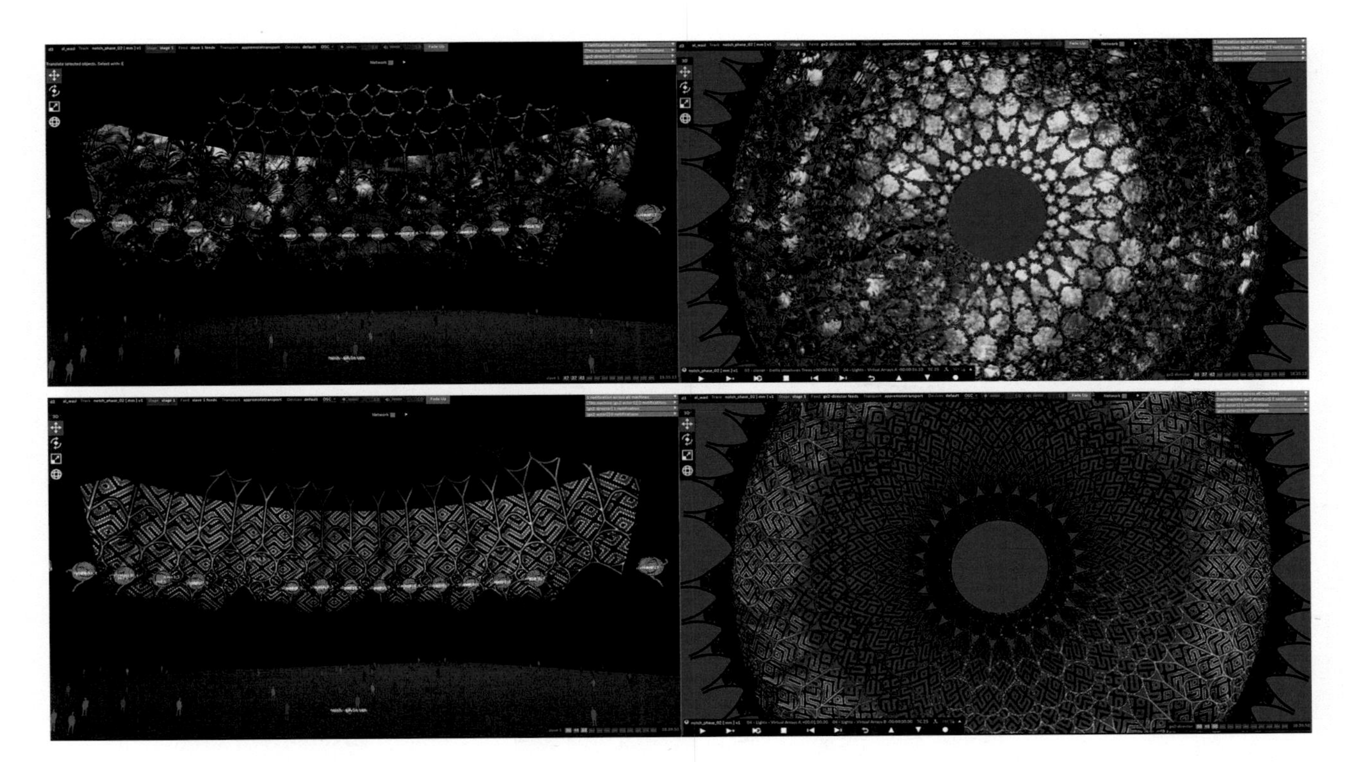

Figure 12.12 Real-time disguise visualization. Right-hand side images show the low- visualization of the whole dome from one camera resulting in a low-resolution quality of the graphics as it renders all the Notch graphics from one server. Left-hand side images show the view of an Actor server that renders the graphics in full-resolution. Source: Image Credit: Bild Studios.

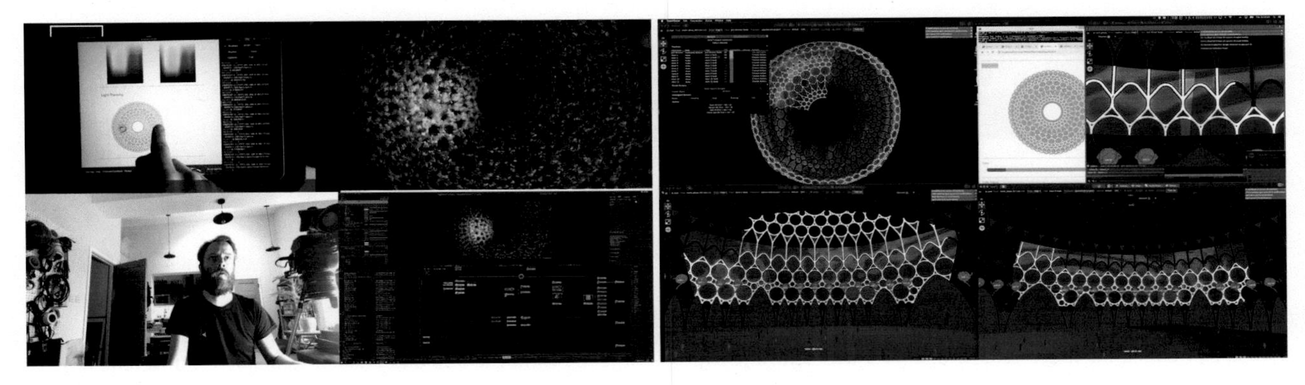

Figure 12.13 Real-time disguise visualization over Zoom calls show interactive app input, a single camera showing low-resolution graphics of the whole dome, as well as full-resolution visualizations from two Actor servers. Source: Image Credit: Bild Studios.

its home screen made up of a grid of thumbnails where the user could select an interactive treatment from. Once the user has selected a specific look its specific controls open up in a new window (see Figure 12.14).

In the backend, a full CMS (Content Management System) editor system was developed allowing a user to customize exactly what elements to include for a specific look. For example, most interactive treatments required a finger tracking interface, access to a texture library and a color picker. Other treatments may have only required a few slider and texture inputs. With the CMS system in place a user can simply pick and choose web elements to include and connect them to corresponding properties in the disguise project (see Figure 12.15).

Figure 12.14 Finalized UI design of the iPad app. Source: Image Credit: Bild Studios.

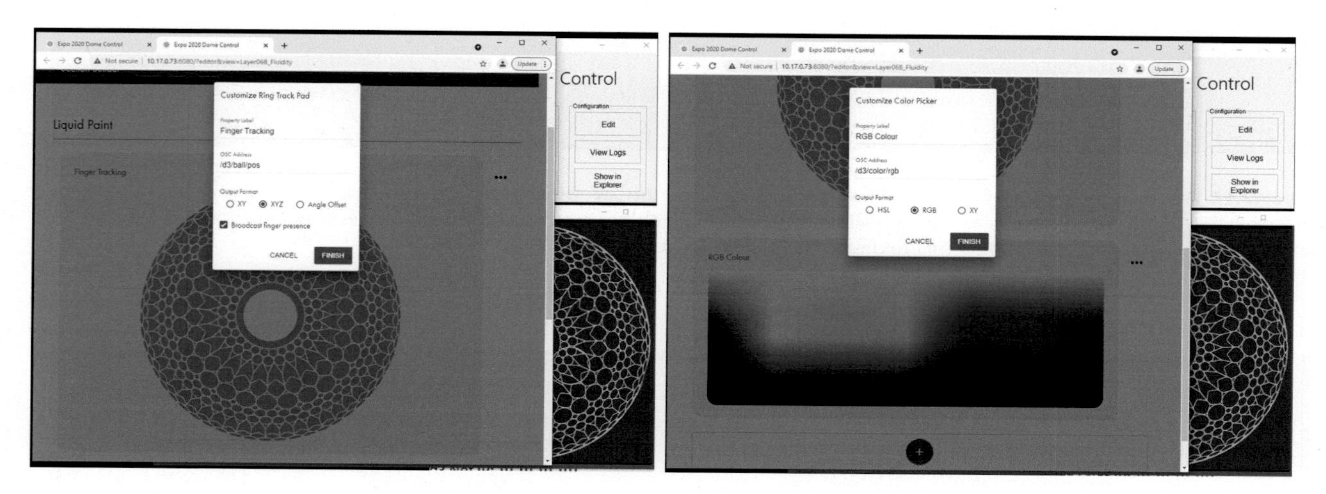

Figure 12.15 These images show the backend CMS of the app allowing the user to add and customize elements required for a specific look. Source: Image Credit: Bild Studios.

With simple actions, such as jumping to another interactive treatment or changing slider values, OSC data was sent directly from the iPad to disguise that then jumped the playhead on the disguise timeline and updated property values accordingly. As discussed earlier, these property values would get input to the Notch block via disguise which then would render the real-time visuals correctly.

Disguise's new Indirections feature allowed video and image content that is stored in a disguise project to be accessed and switched out on the fly from the iPad. Indirections also allowed images to automatically be pushed to the iPad and be represented as thumbnails, with no need to update the app itself.

With this system in place the user could change the texture directly from the app and choose to map it in different ways on the dome. For instance, one interactive look allowed the user to pick up to 10 different textures from the app and select in what pattern the flags should be distributed on the dome (see Figure 12.16). The speed of the flags and adjusting the lighting and shading overall could also be added.

Importantly, to achieve the same effect with pre-rendered content would have involved hours of rendering, most likely with continuous tests onsite until the visuals looked good. With the app, the content could be updated and changed on the fly directly in the dome. This not only avoids time-consuming content rendering but also the use of precious hard drive space.

Figure 12.16 The user can change the texture directly from the app and choose to map it in different ways on the dome. Source: Image Credit: Bild Studios.

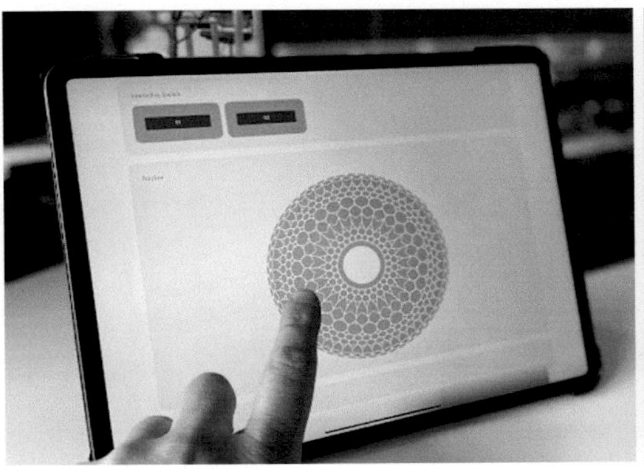

Figure 12.17 As the user tracks his/her fingers on the interface, the coordinates are captured in real time and represented as OSC data. Source: Image Credit: Bild Studios.

Last but not least, the finger tracking functionality of the app allows the user to track his/her fingers on a flattened image version of the dome (see Figure 12.17). These movements get captured, in real time, as OSC coordinates which are sent to Notch, via a web server, then to disguise, and rendered as visuals directly onto the dome. For example, gigantic particle trails or massive virtual searchlights follow the movement of the fingers.

Interactive Audio Setup

In addition, trajectories of sound effects were generated based on finger coordinates of the user and mapped to the 27 speakers inside the space. With the touch of a finger, an additional layer of audio would be mapped to the closest speakers to the visuals generated from the finger movement – creating a full immersive 360 audio-visual experience.

To clarify, the sound itself wasn't generated using a procedural sound algorithm but the *movement* of the audio is. In other words, a pre-created mono-channeled sound file was mapped onto the closest speaker based on finger movements of the user.

This was achieved by sending the same OSC coordinates as sent to Notch to the L'Acoustics L-ISA Controller that then mapped the audio to the correct speaker. In addition, each Notch look had a seven-channel underlying audio track, representing the ambiance mood of each treatment, that got triggered as a new look started (see Figure 12.18).

Right image: the interface of L-isa controlling the L'Acoustics sound system. Image Credit: Bild Studios.

See video of the L-isa interface controlling the L'Acoustics sound system at the companion web page for this chapter rtv-book.com/chapter12

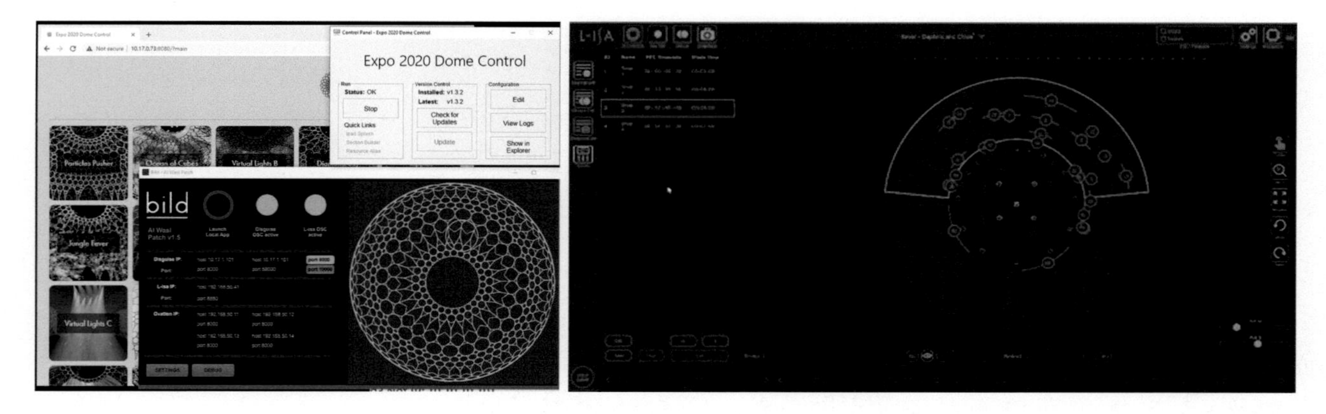

Figure 12.18 The image on the left shows the interface of the web server that communicates with disguise and the audio system L-isa, making sure that the OSC coordinates trigger Notch to render the interactive visuals and audio is being played back at the correct location.

Figure 12.19 Bild team at work on the Al Wasl Interactive Experience. Source: Image Credit: Bild Studios.

Phase 3 – Creative Development

With a technical framework in place, Phase 3 emphasized on creative development and took all the learnings from early stages of the project to craft visuals that both impressed as well as technically performed (see Figure 12.19).

With over 40 different treatments previously explored the team narrowed it down to 20 unique sets of experiences, including responsive virtual lighting effects, huge particle systems, organically growing foliage, dynamically swapped out flags, morphing liquid effects, and many more. Using the previz processes the team developed the visuals and the app-functionality was demonstrated on a weekly basis to show progress.

In parallel, the team also started working to create the final soundscape that would accompany each look. Partnering with London-based sound designer Gareth Fry, each of the 20 looks was given an unique sound design made up of the seven-channel background track and the additional mono-channel which could be embellished via the app controls.

Importantly, a template pipeline was also built allowing the client themselves to build their own unique treatments. As a result, a guide including content specifications, naming conventions, and a user guide for configuring the app was created.

Phase 4 – Onsite Installation

With the significant amount of pre-testing and verification that took place during pre-production the technical installation proved to be a smooth and successful process. The server-performance matched the performance and all servers synced well – as expected.

Not surprisingly, a significant time was spent on color-correcting the graphics, as certain colors appeared a lot more

Figure 12.20 Bild team working with Amna Abulhoul, Expo City ECD, during the final phase of install. Source: Image Credit: Bild Studios.

vibrant than others. In addition, despite a sophisticated previz, certain graphical elements simply had to be refined to appear bigger and bolder to maximize the interactive impact.

While the team expected the audio to enhance the interactive experience, it came as a great surprise what an *extraordinary* layer it added. When interactive visuals followed the finger movement, with accompanying audio traveling through, it truly created a jaw-dropping synchronized experience – nothing else than what we can expect from the world's largest interactive installation (see Figure 12.20).

Thoughts by David Bajt, Co-Founder of Bild and Principal Consultant for Al Wasl Expo Projects

> *"We have 173 shows to create, a new show every day. We will really have an issue creating new and unique content every night and we are really concerned about hard drive space. What do you think, can we run real-time content on the dome? Perhaps we can create an amazing interactive experience too if that works."*

Those were the words from Amna Abulhoul, Expo City Dubai Executive Creative Director, asking me what the possibilities are.

I was excited – it's exactly the type of question we like; a question which I wouldn't have an immediate answer for but we could see the components needed laid out in front of us and an invitation for creative and technical collaboration. All we have to do is to assemble the pieces, right? We knew it would involve creating something new and unique and that would require a huge amount of R&D and testing. If it would work it would turn into something very special. And big. *Very* big.

Once we started and saw the first positive test results from our office setup I knew that this actually could turn into something real. Each step of the way proved more exciting. About half of the first looks we tried actually performed and synced well. Early tests of our iPad app, still running as a web page on a laptop and with emulated finger action, generated big smiles among the team members. When we added disguise's Indirections to swap out textures on the dome, we could see the full picture in more clarity. *This could actually work.*

Once installed onsite and where we could see the excitement in people's eyes when playing with the app, it gave me a huge sense of accomplishment. But to be honest; for me it's always about the ride. To see the puzzle pieces slowly coming together and seeing creative opportunities arise during the journey is the big thrill. The thrill which makes you excited to continue onto the next challenging project.

Thoughts by Expo City Dubai Executive Creative Director, Amna Abulhoul

I come from an animation background, and I myself am an animator. I love doing all the animation work until the step of rendering, simply because it takes so much time. That doesn't work well for me as I like changing and adjusting the graphics to the very last minute and I want to be able to maintain the flexibility all the way to showtime.

What's really interesting and what triggered the idea for creating an interactive application for the Al Wasl Plaza, the largest projection dome in the world, was therefore the requests we got for making graphics *quickly*, even for a show tomorrow. People not experienced in the creative and technical world don't understand the need for rendering time, despite the enormous 27k x 6k resolution of the dome.

I knew that technology is constantly evolving, year by year, and therefore I wanted to explore if there was an opportunity to turn the dome into an interactive solution. With such a solution in place we could adjust graphics on the fly but we would also save an enormous amount of hard drive space and the time for content ingestion.

Even more; if we can create a real-time graphics solution, what's stopping us from letting people interact with the dome? I've seen interactive video installations before but would it really be possible to turn the Al Wasl dome, the biggest and most complex video installation in the world, to an actual interactive piece of art?

When the right minds connect, as Expo is connecting minds together, I knew from day one that the only people who would be able to do this are the same people who already produced the super efficient content pipelines for the Al Wasl dome; Bild Studios.

As we worked through the project stages of the interactive experience, the relationship with Bild was really interesting. I felt we were one team – one mission – one goal. It was a mutual experience where I would listen more than giving instructions to them. I personally felt that I learnt a lot from Bild. I felt I entered a course, letting me understand more about technology and thereby the potential of Al Wasl.

I think the real-time technology represents a kickoff of what we're about to see more of and I can't wait to see how Al Wasl develops throughout the years using the technology Bild has shown us.

Written by David Bajt, Co-Founder of Bild Studios and principal consultant for "Al Wasl – The Interactive Experience."

Project Credits
Bild Studios Ltd
www.bildstudios.com

Bild Studios blend technology and creative production, specializing in the realization of complex video installations for live experiences, Virtual Production for film and broadcast, and permanent media installations. Using cutting-edge media technologies, combined with visual design and production management, their work spans the globe.

Bild's projects usually involve extremely large video canvases, often interactive, supported by hundreds of projectors and huge LED screens that require advanced content/VFX pipelines, playback solutions, and real-time rendering systems.

Being renowned specialists in large-scale live visual experience and Virtual Production, they believe that the unique fusion of real-time VFX, Virtual Production, web 3.0 technology, and live event technologies provides a "technology sandpit" that will revolutionize the way the entertainment industry and other creative disciplines develop.

Bild have also built our Virtual Production Facility, MARS Volume which is an ever-developing production space where virtual 3D environments are mapped onto our LED Volume, enabling actors, and practical sets, to blend into photo-real environments.

Project name: Al Wasl – The Interactive Experience
Project location: Dubai, United Arab Emirates
Client: Expo City Dubai
Year: 2022

Bild project team:
Executive Producer and Principal Consultant: David Bajt
Senior Producer: Lauren Rogers
Technical Director: Jamie Sunter
Project Engineer: Rury Nelson
Software Developer: David Kanekanian
Creative Notch Lead: Lewis Kyle White
Notch Artist: Marco Martignone
Notch Artist: Michael Wilson

Notch Artist: Dave Ferner
Notch Artist: Nick Diacre
Sound Design: Gareth Fry

Expo project team:
Executive Creative Director: Amna Abulhoul
Senior Creative Manager: Karl Knight
Senior Creative Technologist: Charles Draper
Senior Manager – Motion Designer/Animator: Jaanus Vann
Senior Manager Broadcast and Projection: Toomas Vann

SECTION 3

Strategies Using Real-Time Content

CHAPTER 13

Team Leadership and Structure

The Video Umbrella

The Video Department includes a wide number of roles and responsibilities. Whether you are working on a theatrical presentation or a film shoot, job titles will shift and job positions will overlap in different ways, but the team must meet the same essential goal: provide the right real-time content down the correct signal path at the correct moment in time. Clear team organizational structure and tasks are essential for good communication both within the video department and with the larger production team. Let's review a list of core video team roles and responsibilities for a live television broadcast. Many of these roles have corresponding positions in film production, installation, and live events that will be noted.

Content Creation

The content creation team makes video files and/or provides real-time assets that will get processed into video signals.

Within the content creation team:

- Creatives: usually designers and creative directors
- Producers: manage technology, schedules, quality control, budget, asset delivery, and project assignments
- Animators, Technologists, and Creators: build files and assets for use
- Engineers and Technologists: support the infrastructure to create assets
- Associates: fill in all the spaces in between

In film, content assets will come from the VAD (Virtual Art Department) while in a theater content creation is the responsibility of the projection design department. The more complex the project and budget, the more granular the roles related to content creation will become. Similarly, in small projects, one artist might be solely responsible for developing and delivering content.

Media Operations

The media operations team manages the video files or real-time assets on a production site.

Within the media operations team:

- Producers: run video productions team onsite, lead client and inter-departmental communication
- Programmers and Operators: manage the control of playback or processing technology that delivers assets to video signals
- Engineers: design and maintain the playback and real-time systems delivering video signals, the data networks, and the synchronization systems like genlock and timecode
- Integration Engineers: facilitate interaction with partner departments to monitor shared infrastructure like camera tracking, 3D data collection, and signal delivery
- Content Managers: direct asset collecting and file versioning
- Project Managers: support the team and maintain task timelines

I started using the term "Media Operations" team in the late 2010s. I have not found any term consistently used to describe the team of people responsible for delivering video content to screen. As previously described, "Video Department" serves as a catch-all for a multitude of roles and responsibilities. In theater, the projection design team is usually separate from the video department. Outside a traditional theater, this delineation is less clear.

The term "brain bar" is used in film production to refer to the team handling VAD assets and the game engine rendering system. Many find this term undesirable but a suitable alternative has not taken hold. I like terms like MediaOps or VADOps, borrowing from the software development DevOps. Our workflows and process management have many similar demands.

DOI: 10.4324/9781003206491-16

Further confusing the taxonomy of our industry, the Ops team role can be found as part of the content creation team, or the system engineering team, depending on the project. Wherever these roles land in the department structure, they tend to be the center of project communication and a critical nexus point of shared information.

System Engineering

The system engineering team configures and manages video signal processing and delivery to either a screen or broadcast destination

Within the system engineering team:

- Project Managers: organize and direct the engineering team
- Engineers: plan system technology and redundancy
- LED Technicians and Projectionists: manage physical gear
- Technology Specialists: handle specific real-time tools related to camera tracking, background keying, and 3D data collection

Broadcast Engineering

The broadcast engineering team controls live video signal collection and delivery to a live audience.

Within the broadcast engineering team:

- The Camera Department: responsible for shooting and cutting the live feed
- Directors and Producers: monitor the live feed and direct the line cut
- Engineers: secure signal integrity and delivery
- Utilities: support the technology and team

Broadcast engineering will include camera departments in a conference or live audience event with IMAG. In these cases, live broadcast is localized for live audience viewing and not delivered to audiences outside the event venue. On a film set, I would look to the camera department to manage related information.

Video Team Leadership

Content creation, media operations, system and broadcast engineering are all part of a larger video production ecosystem, but each exists as its own video department. Each team operates as a discrete unit, sometimes unaware of other video departments' needs or responsibilities. Because of the complexity of this fast-growing discipline, many teams may be designated or considered "video" but, in truth, only run a portion of the overall video delivered for a given production.

Remember that broadcast engineering existed long before there was a video content creation team and the content creation team arrived to create on-air graphics, well before video screens of any kind were part of an entertainment production design. We are still learning how to communicate with and support each other.

Real-time content production creates even more interdependencies between video departments. It is critical we are constantly reviewing our communication to our partner video teams and how it affects everyone under the umbrella of the video production department.

Consider an LED stage with a surround, floor and ceiling made up of 1.8mm LED tile and two cameras. The content creation team has multiple design and delivery requirements to consider. Information about the rendering capabilities of the real-time content platform will inform the amount of detail that can be managed in the content design. Storyboards or previz files will outline the shot, determining how much of the world needs imagery to fill the background. Meanwhile the camera department, and the screens and system engineering teams will need to review frame rates and shutter angles with the refresh rate of the LED screen. If the stage is used for a live broadcast, the AR elements of virtual scenic extension will require time and testing. Communication and planning are key at every step.

With so many key details to cover across so many facets of a video department, who is in charge? I would argue this question is one of the key factors limiting video production team growth in the entertainment production world. Senior project leadership does not see video as a cohesive department because we are not bound under our own clear leadership structure. Instead, we assemble on a project as disconnected video fiefdoms competing for time, budget, and resources. We need a different paradigm.

Good leadership starts by bringing the video disciplines together under clear direction from a singular team producer that can facilitate communication across all video sub-departments. To achieve this, we must first establish best practices and then define a team producer capable of supporting the success of all aspects of video. Unfortunately, this method is uncommon. We as a video production community have to insist on a clearer leadership structure.

What might that look like? For real-time content projects, I recommend a Virtual Production Director or Screens Producer who serves as a single executive leader to the video teams. Some projects might be complex enough to need both, especially if the real-time component is only part of the production goal.

The production team members who oversee all the varying aspects of video on a project support and are, in turn, supported by that executive leader. I like to think of this as pixel support from design to display. Good pixel stewardship by an executive leader supports creative teams with content design, operations teams managing content, and engineering teams delivering content to screen, on a stage, or to a viewer's home.

> Who is your video team leader? Who oversees real-time video production from design to display?

The video team will vary depending on the medium: film shoot, a webcast for a corporate event, a concert tour, or a broadcast award show. For every kind of production team, leadership cannot concern itself with the success of only a single facet of video delivery. Someone must supervise communication across all the video teams and their production partners, and take accountability for holistic delivery within the overall production.

As you begin working in real-time video, consider your scope of responsibility and how it affects members of the different video teams. Are you handling 3D models? Have you provided a clear assessment of model size to the engineers responsible for detailing the technical spec of the servers hosting those models? Are you in the camera department working on an LED stage? Have you discussed the implications of preferred frame rates with the screen engineers? Think beyond your own success and your work will get easier and your success more assured: the entire video team's success is your success. Communicate effectively internally and across all aspects of video departments and the video team will collaborate more effectively with external production departments.

Partner Teams

On a production, several partner teams work with and around the video department. Each of these teams benefits from clear information and direction from a unified video department. Most common and prominent among video's peer teams are Scenic Design, Lighting Design, and Audio.

Scenic Design/Art Department/Virtual Art Department

Whether using physical scenery or virtual scenery, the scenic design team will include a designer, art directors, draftspeople, and other supporting roles. Scenic design is an important ally of the video team now that a large part of the available production space for design includes video screens or virtual elements. Set designers have unique skills for planning performance spaces and understanding how space is used by performers and experienced by the audience. Competent set designers offer great insights into how to use screens effectively in performance.

In many production environments, the scenic team maintains responsibility for the vendors that supply video screens on a production. They are your resource for the technical specifications critical to building a workflow.

With Virtual Production, scenic design has become an extension of content production, as the set is partially or entirely digitally created. Design assets from the scenic department should flow easily between art directors and content creators. In film, the Virtual Art Department effectively leads content design as the team responsible for defining the look of the background.

Lighting Design

The video department and the lighting design team interact throughout every production. Establishing color temperature to balance the value of white for cameras often falls to the lighting team. This will significantly affect the color of video screen content. Discussing stage looks and colors with the lighting team is critical for a good-looking presentation.

As screens dominate stage presentations, lighting scope has reduced on many productions. And on a film LED Volume, screens often become the primary lighting source. I expect a real-time lighting specialization to emerge soon to track the screen driven "natural" lighting and special sources that amplify those effects.

Real-time content tools present an interesting opportunity for lighting-driven interactive video controls. As with scenery, we can extend lighting rigs virtually into video screens and even control them with DMX lighting protocol.

Audio

When dealing with the audio department, the video team often looks for timecode to enable proper synching and audio level management for the video files with sound. For real-time content, audio interactivity is a regularly used feature. Sound used for content generation requires more audio department interaction.

Microphones, audio analyzers and other sound gear may interpret the sound for image manipulation and generation. This usually requires someone on the video team familiar with sound and sound technology.

Clients of the Video Team

Beyond the more immediate circle of video peer or partner teams, we see other departments that we can view as clients of the video team. These teams define expectations and set the goals for wider artistic and financial goals for the production.

Director

Film, television, and theater all have a director role on the team. While collaboration with other design teams will vary depending on the production, a director's vision defines what the audience will see on the video screens or through the camera lens. How does a director plan staging or camera shots if the visuals produced depend on the staging or camera position? How is the content design shared when the result is generated live on stage?

Storyboarding and previsualization can help. For example, in Virtual Production, we can build the entire scene in 3D along with a virtual camera that responds to live input via keyboard or mouse. Based on the result, the director can ask for changes or enhancements before the actual shoot when changes become resource intensive and expensive. Together with the video team, the director can plan camera shots with previsualization while generating the resulting content for that shot.

In a live event or installation, we can provide a virtual previsualization of the full production. Simulation of the real-time content will expose creative misunderstandings or potential production challenges. It is better to find these issues and correct them prior to production when hours are more affordable and deadlines aren't looming.

Producers

Previsualization not only supports the director but also provides immense value to the Creative and Executive Producers. As a video department, we must communicate plans to our partner teams, but we can also facilitate client communication with our clients using 3D previsualization. During production planning, we must also clearly communicate the effects of real-time content creation on budget and schedule. Many producers rely on template outlines for their cost and time expectations. Producers regularly underestimate the budgetary and timing impacts of Virtual Production.

Our challenge is to prepare producers for the reality of VP without destroying their appetite for creative innovation. We need to inform without making this work seem too difficult to manage; to set expectations while not scaring producers off of working with these tools later when they become more affordable.

We will dive into budgeting in more detail in Chapter 15. First, let's learn about how a well-structured team delivered the 2021 MTV Video Music Awards.

Case Study – Visual Noise Creative – MTV VMAs 2021

The screens team for The MTV Video Music Awards included the following team members who were responsible for the AR visual elements:

- Screens Producer
  - Responsibilities included staffing up the media operations team
  - For this project, the screens producer focused on a holistic creative approach to the unifying screens, projection, and AR content in a cohesive visual statement for the performances and show altogether
- Technical Producer
  - Responsibilities included the oversight and integration of the AR-specific system into the overall screens system
  - Engineering of an LED playback system and a projection playback system
- Content Studio
  - The studio authored the AR content in Notch and delivered Notch blocks to our content manager
- Content Manager
  - The manager received, QC'd, and updated files and Notch blocks
- AR Engineer
  - AR technical supervision
- AR programmer
  - AR elements were sequenced onto the disguise timeline

Figure 13.1 Projection Mapped Moon Person with AR Graphic overlay. Source: Image Credit: VNC for the MTV VMAs.

For the real-time content to be successful, the AR space needs to be calibrated to the real world space. On this show we utilized two tracked cameras – a technocrane, and a spydercam. The alignment process used Stype calibration into disguise. Stype performed lens calibrations on both cameras, used Stype encoders for feedback from the technocrane, and interpreted winch data from the spydercam.

The creative intent of the real-time content was to combine synchronized projection and LED elements captured on camera, with a 3D composite of AR elements. Visually, this created unique and immersive show moments focused around awards show patterns. Awards categories such as "Artist of the Year" displayed on top of the Moon Person, a 40′ 3D projection mapped landmark of the production (see Figure 13.1). AR was also used during performance moments to bring scenic elements and musicality into all areas of the arena space.

A key component on a show of this scale and speed was appropriately dividing and dedicating technical resources to achieve success at a broadcast pace. This meant distributing the server workload between four unique disguise systems – LED, Projection, AR, and Band Adds. Additional video gear brought in by performers can be unpredictable in scope and decided well after the system is engineered. We've learned to invest in gear that might get minimally used to protect the primary show systems from unplanned processing demands by adding the Band Add system. Similarly, additional workload is put on the programmers and the decision was made to keep these systems discretely controlled. This meant dedicating a programmer to AR.

One challenge faced was how to make four technically discrete systems (AR, projection, LED, Band Adds) appear cohesive and unified creatively. Many show elements involving AR had a corresponding projection component to create one immersive display of content. Each element was being played back uniquely by its respective system, with one programmer handling LED and projection via Sockpuppet, and a dedicated AR programmer utilizing the disguise timeline.

Each system received two sources of timecode – EVS playback and show Protools. EVS was utilized for video package playback during awards moments, while Protools was used for music playback for performances (see Figure 13.2 for an AR bear used in a performance by Ozuna). The EVS system supplied timecode that was used to synchronize projection and AR to the awards packages. The Protools playback timecode was used during performances to synchronize AR and LED elements. The

Figure 13.2 Ozuna performs with an AR teddy bear. Source: Image Credit: VNC for the MTV VMAs.

timecode source was switched upstream of the disguise system, to provide the correct sync source when needed.

Go to rtv-book.com/chapter13 to see the Ozuna performance with it's AR components

Our AR engine for this production was Notch running natively on the disguise servers. The Notch block for house looks utilized both pre-rendered content, as well as live inputs for playback lines from the broadcast truck. This allowed for packages to play inside the AR elements, triggered by the truck's EVS playback system. The show's director would cue these elements to play back in AR, to be shot on camera, then cut the package full screen. This required a good deal of coordination between audio and video to ensure sync was accurate between the source and all destinations.

While the systems were engineered with the lowest latency possible between components, there is still inherent delay in the chain of capture, tracking, and output. Sync between elements is of critical importance, and our team worked in tandem with the tech management and truck engineering team to ensure the entirety of the camera system was engineered to account for the inherent latency of the AR system.

Adapting established workflows and technologies was the key to success. It was the responsibility of the screens team to push the creative capabilities while being mindful of the process of live broadcast. A staple of live broadcast production is reliability and speed. For example, the ability to ideate on the AR files quickly so as to not hinder rehearsal times was integral to rehearsal. Files had to be pushed and QC'd ahead of time so free moments with the director could be capitalized to push the house AR looks forward. The entire team of content studios, content managers, and producers had to be working in the background to make sure we never fell behind and presented fully prepared looks during scheduled rehearsals.

Employing AR technology to push creative boundaries in live broadcast comes with the necessity of a stable and dependable system. We only get one chance to achieve success with these moments. With so many technical pieces working in tandem, our team spent a good deal of time planning to ensure we were deploying personnel and systems to deliver a 3 hour live broadcast with no hiccups. The research and development of the team leaned heavily on accounting for bandwidth and headroom in the systems, reconciling timing discrepancies and latency across all components, and furthering workflows for playback, asset management, and integration. See Figure 13.3 for an example of the result during an awards presentation.

Figure 13.3 Artist of the Year AR Look. Source: Image Credit: VNC for the MTV VMAs.

In the end, the final product can only be as good as the alignment of the two worlds which is a function of accurate data. We started with a 3D scan of the room and scenic elements. We calibrated the AR cameras to that scan. The movements of the spydercam through its flight envelope, over time, would lead to small mis-alignments, requiring the camera's position to be re-homed to bring the worlds back into alignment.

This was difficult to achieve during the show as the spydercam is almost constantly in motion, keeping up with the action of the show. Due to the realities of unexpected drifts in position feedback, our engineer modified and compensated live by moving the 3D scan around during the show. By slightly adjusting position fields, he was able to keep the virtual world in alignment during some scripted camera movements with the spydercam.

We invested considerable time in creating a team structure that could support the production creative demands, support AR for live broadcast, and keep the people on the team under realistic work pressure. The trust built with our MTV producers over the years means we can have sensible conversations about budget and time to meet cutting edge creative goals. We've created a trusted partnership that is creatively satisfying and a good work environment.

By Emily Zamber for Visual Noise Creative
www.visualnoise.net/

Visual Noise Creative LLC is an entertainment, design, and production firm that believes in the power of storytelling through technology. We believe that good design draws in audiences with a strong narrative, a firm point of view, and that the collaborative process produces design on the highest level. Visual Noise is adept in shepherding projects from the earliest stages through realization, all with a high value being placed on process. In other words, we think it should be fun!

Project name: MTV Video Music Awards
Project location: Los Angeles
Year: 2021

Project team:
Screens Producer: Trevor Stirlin Burk
Technical Producer: Zak Haywood
Screens Programmer: Kirk J Miller
AR Programmer: Zach Peletz
Content Director: Kerstin Hovland
Virtual Spike Mark: Joe Bay
disguise Projection Mapper: Benjamin Keightley
disguise Projection Mapper: Rodd McLaughlin

Server Engineer: Tim Nauss
Server Engineer: Ben Roy
AR Content Manager: Ann Slote
Account Representative: Martin Wickman
Spyder Engineer: Jason Spencer
Screens Project Manager: Emily Zamber

Client:
MTV-Viacom Media Networks
Executive Producer, Music and Talent: Bruce Gillmer
Executive in Charge, VP, Production: Alicia Portugal
Den of Thieves Executive Producer: Jesse Ignjatovic
Den of Thieves Co-Executive Producer: Barb Bialkowski

CHAPTER 14
Inter-Department Communication

A successful production process depends on efficiency. When one department runs into delays or issues, it can block the work of other departments or the entire production. To create those efficiencies, each department must anticipate requirements for accomplishing their goals and prepare alternate actions should their best path forward get disrupted. Disruption can include internal issues, external hold ups, unexpected resource gaps, or a sudden creative redirection.

We joke about all night redesigns to appease a disgruntled performing artist or commiserate over long days solving technical lapses of other departments. In truth, such stories make us sound heroic for the wrong reasons. Working excessively long hours should occur extremely rarely. Hopefully, enthusiastic pursuit of our own creative desires fuels our occasional 100+ hour work weeks, not the desire for a bigger paycheck and certainly not to assuage a client's creative temper tantrum.

What impressive stories of human effort cover up is weak communication. Clearly explaining a creative vision is hard work. Blue sky ideas need to become 3D models, spreadsheets, and gear lists. Someone getting paid real money for their creative vision needs to turn inspiration into mundane paperwork that communicates that vision to the broadest possible team. Production work is a collaborative effort and supportive collaboration is about good communication.

Turning art into documentation doesn't corrupt the vision. On the contrary, inspiring and coordinating a group of people to take every step, inspired and mundane, to realize that vision is immensely rewarding. Each job position invites different creative expression in service to the larger creative goal. For one person, it might mean organizing cues on a video timeline. For people like me, it's the love of a well laid out and thorough spreadsheet. You can't convince me an excellent Google Sheet doesn't require artistic expression.

Production work is ultimately about community. We come together and set about meeting a goal put forward by a Creative Director or Executive Producer. We establish the responsibilities and expectations of the various production roles in order to simplify communication. Work with the same group long enough and communication seems to disappear. When teams know one another's roles and responsibilities well enough, the production process appears effortless.

The challenge a creative video department faces is that we are a new paradigm, and real-time content creation is in its infancy relative to other production departments. There is no communication shorthand in a discipline that does not have a common workflow. Certainly, we have reviewed code-based generative content creation going back to the first computers of the 1960s, but Virtual Production only became a commonly used term to the wider entertainment production community in 2021. Real-time content creation came into use around 2017. We have only just begun to learn how to communicate with each other within a video department, let alone how to explain our processes and requirements to partner departments.

Video teams need to overcompensate until real-time content creation becomes better understood in the production world. Iteration and experience are the only way anyone learns a brand new production medium. We must continue to bring our clients on this journey by transparently outlining our goals and process on every production until we begin repeating ourselves. Every production is an opportunity to educate, those innovating new ways of working and new ways of seeing even more so.

Crafting Good Collaborative Relationships

Good collaborative relationships are based on trust and competency. Even if you are doing something for the first time, competency in prior work or trust established through personal recommendations is essential. This is the starting point for

DOI: 10.4324/9781003206491-17

building a production partnership. We are stewards of complex technology that is not well understood. To navigate a discipline that is hard to explain requires thoughtful and patient communication.

You must also know what to communicate and when. Not every client will want to know the minutiae of how you do your job. Your production partners must focus on meeting the challenges of their own project goals. Make it a priority to provide useful information when it is needed. Be clear about what it takes for your team to be successful and what issues are preventing you from getting there. Learn how to deliver this information succinctly. Let's lay out key points on how to structure communication.

Step 1: Understand the Task in Front of You and/or Your Team

When you are getting started in the creative video field, it's easy to place all your trust in the person responsible for hiring. The expectation is that someone has reached out because they understand what your skills are, what your job is, how many people are needed, and how much time is required to succeed. Unfortunately that is not always the case. Sometimes a hiring call comes from an experienced VP producer who has a particular task in mind and a clear understanding of the job they need filled. Or the call may come from a production assistant who has never heard of real-time content.

Be prepared to ask questions. There are basic employment questions for an individual that cover, number of days, rate, travel, pre- and post-production time, per diem and accommodation, and expected tools you will supply. These points will become more complex if you are tasked to build your own team. Then there are the roles and responsibilities of the job or project you have been contacted for. What are the expectations of your participation? What is the project goal? What are you counted upon to deliver?

Ask yourself if you have received clear information regarding the job. You may have been contacted because of a particular skill you have, and not by someone who understands how that skill will be applied. This is often where expectations are miscommunicated. Take a little time to collect information about the project and what is expected of you before you accept the position. This can be achieved in one phone call, or you might need a day or two so that you can review available documentation and confirm if you are right for the project. Expect clarity early on so that you can be successful in your approach to your work.

Step 2: Repeat Your Understanding of the Task Back to Your Client

If you are leading a team or working within the team, at some point you will have enough insight into the task ahead to make a summation of the project back to your client or team leader. Take this as an opportunity to not only clarify your understanding of the project goals, but to help your client refine their expectations of your participation. Everyone in the team is responsible to someone higher up the production team leadership. Regular discussion of your responsibilities and tasks might uncover production changes not yet communicated to you. Perhaps a camera has been moved, that resulted in a small scenic design change that impacts a screen size, camera tracking, and content design. It happens. Every point of discussion and review might collect further clarity on the project itself that informs your approach.

I recommend a simple daily check-in of what was accomplished, what your next steps are, and what obstacles you face to get there. This exchange should be simple, concise, and clear.

Step 3: Outline Your Plan; Review Your Plan's Impact on Your Video Partners

As you understand the project goals and your responsibilities within that project, you will make choices on how you intend to complete your tasks. This is often a moment that you can end up making assumptions about what other team members will be responsible for, or understand your approach to meeting your team's responsibilities. Draft your approach and share it with the production team. Check in with other video teams that your work will affect and see what the implications of your choices might be.

Let's say you manage the engineering and maintenance of a media server that hosts a real-time video project file. You are likely going to be interacting with the team creating project files and the team receiving video signals from the real-time platform. After receiving a draft of real-time assets, you see that the project file is very demanding and the server frame rate is below an acceptable result. You can either talk to the content team about simplifying the project file, or you can improve the type or quantity of media servers to support the project file. The latter will increase the number of video signals delivered to the screens engineering team. You cannot make this choice without understanding the effects on your partner video teams. There are time, gear, and budget implications with every option.

Could you have anticipated this issue? Maybe and maybe not. Perhaps the content team's final output exceeded the precise capabilities of the system and polygon count limit for 3D models that you had previously documented and shared. Perhaps you specified a system with media server spare signals that is already approved for budget. You can perfectly outline all the system requirements, but that doesn't ensure that your video partners will understand or follow them. Documenting the system and its effects on your partners will help when these discussions come up later.

Perhaps you have to make choices that have impacts outside the video department. There may be depth cameras or IR receivers that require highly controlled locations to function properly. These locations could impact any other department on the project. There might be specific lighting conditions that are necessary for the cameras and receivers to work properly. Before you commit to a specific technical solution, understand the requirements and the impacts those requirements have on other departments.

Step 4: Discuss Options and Efficiencies to Improve Building the Project

After you have outlined your approach to the production tasks expected of you or your team, you might see there are ways you can improve reaching the goals with slight variations. Based on your own strengths, capabilities, and personal interests, parts of the project will be easier, more interesting or more motivating than others. Perhaps you can make a challenging task fit into one of those categories with small adjustments to the requirements as you understand them.

Often efficiencies are gained through a change in technology. You might have been hired for your experience in one real-time platform, but you see ways another platform will achieve the project goals and save time or budget. However, these types of insights don't have to be limited to your personal effort when building content, cues, or workflow. Sometimes the exact requirements of the end goals are negotiable.

For example, time is a valuable asset. I have often approached a producer with minor adjustments to a creative goal that will save time. As you are reviewing project options with stakeholders, understand which things might be described as "wish list" items. If there is a way to deliver 90% of the goal in 50% of the time by making minor changes, that may be a valuable discussion for everyone involved.

Step 5: Build communication tools to direct team operations

Documentation is an essential communication tool. Clear documents describe your approach to meeting your responsibilities, and will aid any partner teams that must meet those same or related goals. When you outline team workflow, you will detail a process to deliver content, engineer a system, or pre-visualize creative. There are likely many other teams reviewing and learning from the documentation you create. Documents explain, organize, and track results.

Make your documentation easy to digest. If the documentation is too difficult to navigate and makes finding useful information time consuming, it will not get used. Visual documentation is more time-consuming to produce but easier to interact with and understand quickly. If the information is best conveyed through text, maintain a concise, consistent, and logical step-by-step approach.

Share documentation internally and externally and listen to feedback. You can't create a perfect information resource for everyone, but you can find elegant compromises. A team runs best on communication tools that people actually use and know how to navigate. Watch how users of your communication tools interact with and are made successful by those tools and make note of what they miss. Your documentation is always capable of improvement, either on this project or the next.

Step 6: Continue to Engage Video Partners on Their Pathway to Meeting Shared Project Goals

Once a production starts, regular communication is necessary to understand what your partners need from you and to outline what you need from them. Production is a continuous dialog and staying one step ahead of issues will make them easier to solve. When challenges inevitably arise, good communication will limit negative outcomes for you and your team. Regularly engage production partners to ensure mutual understanding of the end goal. When your partner teams feel supported, they will reciprocate when you or your team faces challenges.

There is no perfect science or process to working with real-time content in any production environment. Everyone is learning the best way to work with these tools. Often, this comes with the challenge of unlearning old and formerly very successful behavior. Don't resist learning a new approach; share the new insights you find. What you will come to find obvious in your task will not be obvious to everyone around you. Work with patience and clarity.

In the following case study, we will see how Evoke Studios built an XR Stage for Siemens. In the study, the team will detail how they established what they were delivering for their client and how they intended to deliver the project.

Case Study – EVOKE Studios – Siemens Hannover Messe Digital 2021

Founded in late 2019, Evoke Studios functions at the intersection of creativity, workflow, and bleeding edge technology. The multinational team loves working on complex, high end projects in many different markets and contexts worldwide, specializing in real-time solutions. It's been a challenging experience for the partners launching, building, and nurturing a business rooted largely in live events, throughout an unfolding pandemic. As a result, like many others, the studio has pivoted to providing leading solutions and workflow in Augmented and Extended reality design to great success. Let's take a look at some of the projects, challenges, and solutions created along the way and how they relate to real-time technology.

In October 2020 Evoke Studios started co-development on an Extended Reality solution for Siemens Hannover Messe Digital 2021, the concept of which took the form of a Unity based virtual showroom (see Figure 14.1).

Hannover Messe Digital is an online trade show platform where Siemens interacts with its clients and provides their audience with new product presentations and general company updates. This is accessible worldwide. The Siemens product range for this event has a strong focus on industrial machinery, robots, and precision tools, all of which are large and cumbersome to transport and more difficult to make part of a single event, without a physical trade show booth to visit.

The concept for the event is based on a pre-existing Siemens virtual exhibition application hosted on AWS for web and mobile devices based on the Unity engine. They reimagined their web app as a virtual studio environment for use in broadcast scale XR, making it applicable for use event-wide. The footage coming out of the resulting XR shoots was then shared on the virtual platform through bookable client sessions with support from Siemens personnel.

It was of key importance to the client that textures, colors, and animations of their products would be accurately maintained, and that they would be provided with a responsive workflow during shoot, as the speakers all had busy diaries and assets would not be delivered until right before start of the shoot.

Figure 14.1 Siemens XR Stage. Photo Credit: Chema Menendez.

Using XR and real-time rendering within this project had the following uses:

- It provided flexibility to the camera director in finding engaging shots – pre-rendered content which only works from one perspective would be far less flexible in production.
- It supported the at times less experienced speakers in locating themselves on stage, providing visual feedback for the talent and director – something a green screen solution could not have done so easily.
- Front plate and back plate live positioning and key framing of objects helped with choreography.
- We were able to create large virtual camera moves on the fly, not possible without real-time rendering.
- The render engine platform allowed us to stay native to the manufacturer's workflows, saving time throughout the production process.
- Providing a better end result over using traditional methods of working.
- Previsualization of scenes and verification.

Technologies used:

- Extended Reality
- disguise media server platform
- Unity render engine
- Stype RedSpy tracking solution

> *"The xR environment is so demanding that you will need to keep your knowledge and skills up to date, every day new technologies are being developed and probably you will need to integrate them in your workflow."*
>
> – Chema Menendez

Doing Something First, It Can Be Tricky

Unreal based XR, and to some degree Notch based XR was already more established by the time this production got underway, Unity as a render engine platform had not seen much if any use in disguise enabled XR yet.

If we look at the state of the disguise XR platform at the time, it did aim to fulfill the role of a render engine agnostic platform – and in theory it could – but the Unity plugin was not Virtual Production ready yet at that time. And yet here is a client who desperately requires it, which triggered us to develop a plugin of our own. For the Siemens SPS Dialog which came before this event, the pipeline consisted of translating everything from Unity to Notch, which was far from efficient, but more feasible at the time.

We approached disguise to see if they would support us developing our own plugin for this purpose, and they were open to that. Getting a from-scratch designed Unity plugin to work with disguise, with all production requirements covered, as it turns out, can be tricky. Creating sensible controls and interfacing, key-framing from inside disguise, and running in a stable fashion while every other new disguise build would likely break everything again, was a challenge.

This work resulted in the ability to stay native to the end client's Unity workflows, using their assets more easily, optimizing them for use in real-time, and avoiding having to rebuild everything from scratch for another render engine, which would have been a much more costly and time consuming approach in comparison.

Plan, Plan, Plan!

Due to the topology which makes up a functioning XR solution, there is arguably no such thing as an off the shelf solution for XR (or by extent AR or in-camera VFX). The basis of broadcast scale XR solutions rely, amongst others, on products from multiple manufacturers, their supply chains and cost, availability in rental and geolocation, the considerations and skill of the particular systems integrator and creatives involved in the project (and shhhh, the politics!), overall project requirements, whether it's temporary install or permanent, and not to mention basic circumstance.

Factor in next, all XR solutions rely on software. Most of which at time of writing exist in more or less of a stable state (some manufacturers would tell you they can do XYZ regardless of the stability implications or if it's even capable of delivering XYZ on time at all – marketing being ahead of things could be considered endemic within XR). Project requirements can often function ahead of the state of the software, which both drives innovation and informs expectation management which makes up a potentially large part of your responsibilities. Examples of software interfaces can be seen in Figure 14.2 and Figure 14.3.

Because of this we would recommend that you consider taking the time to properly scope out and plan where you might find

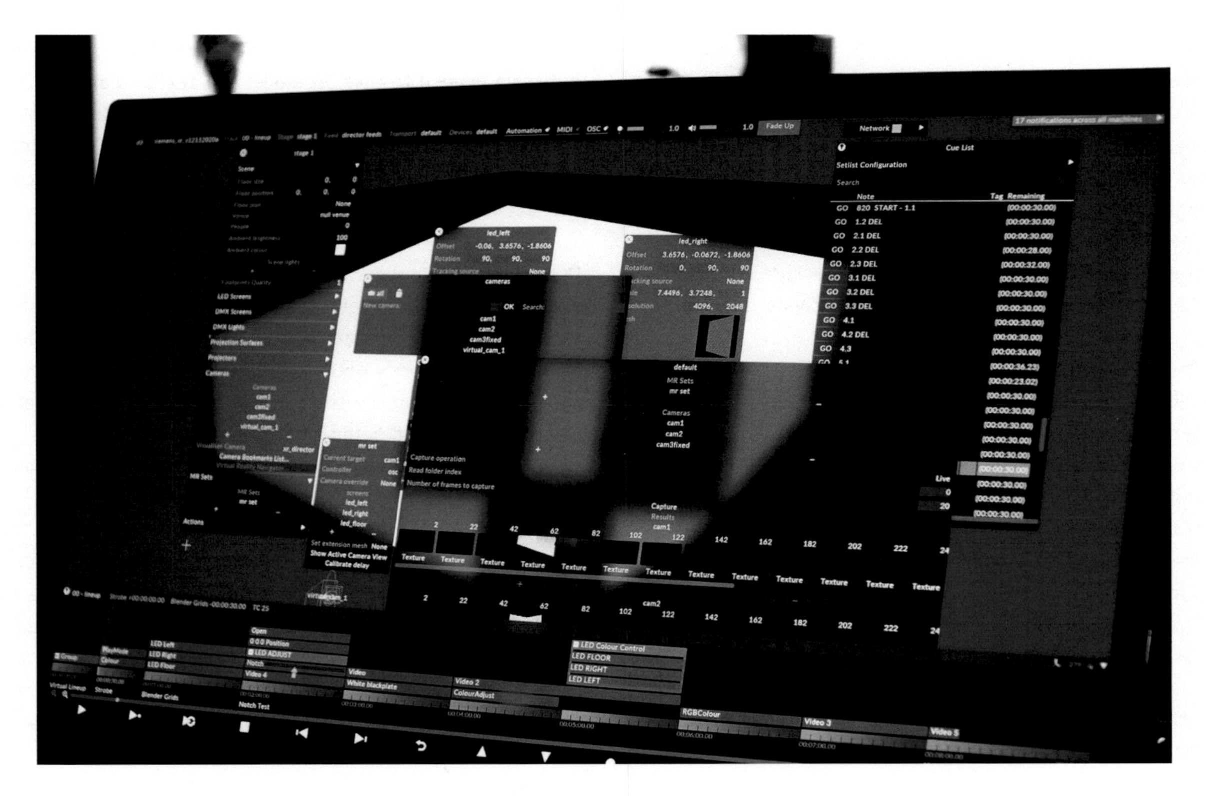

Figure 14.2 Disguise interface in use on Siemens XR Stage. Photo Credit: Chema Menendez.

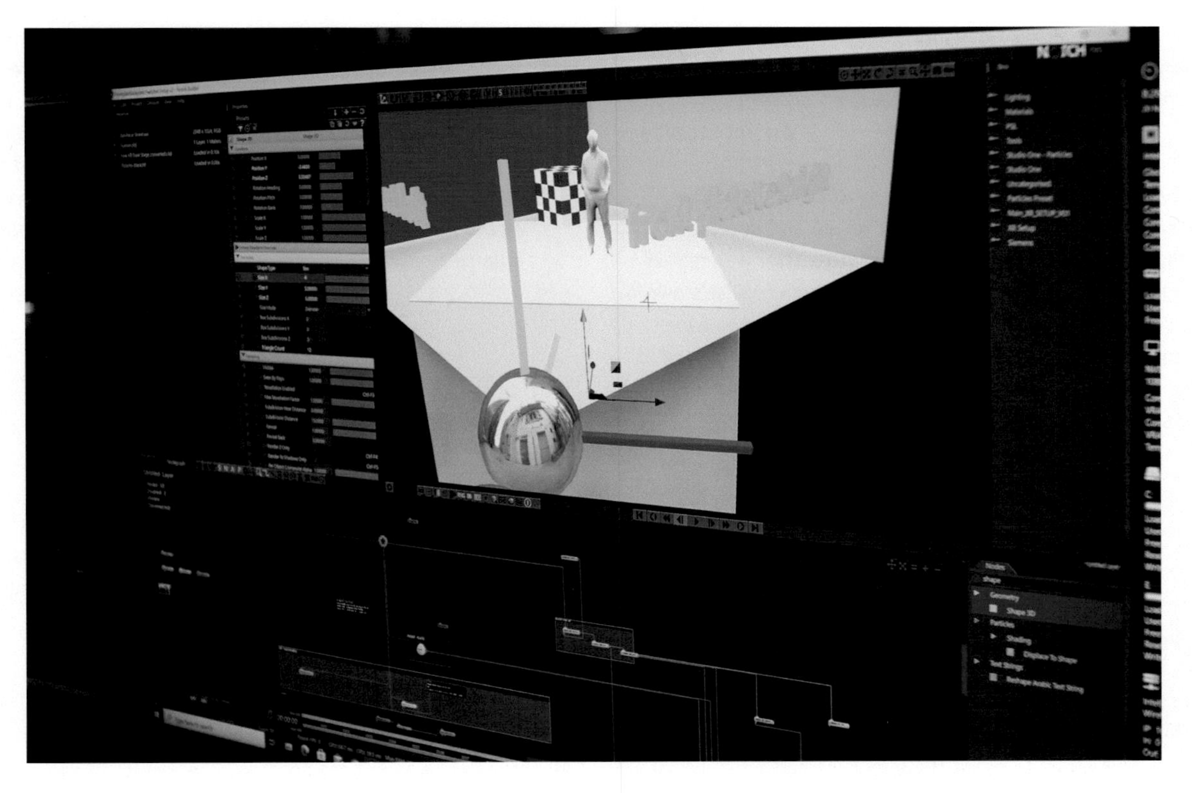

Figure 14.3 Notch interface in use on Siemens XR Stage. Photo Credit: Chema Menendez.

risks within your project and solution; which parts of it are being newly developed, how much time it takes to test and verify (parts of) the solution, are your people skilled in the right areas and available, and how much contingency you wish to plan with. Consider the benefits and drawbacks if you can plan with overtime or staggered crews, and is it more important to be affordable to the client, or to preserve the energy of your crew throughout a project and to mitigate burning people out, because XR takes hours to keep things presenting within milliseconds.

XR is made up of a lot of factors, most of which can now be relatively safely calculated against to ensure on-time delivery. Whether this means calculating milliseconds needed to render a frame, or putting stars up on a ceiling for a tracking system and calibrating a space, these potentially become familiar factors to you soon, depending on your area of work. However, software *can* fail. Technology *can* fail. Set your systems up to be redundant where possible and plan with contingency so you minimize the risk of failure to deliver – do your due diligence and make sure you can feel secure about your approach.

> *"When dealing with real-time graphics and tracking data, the most valuable thing you can give yourself is time. Losing a day on smoothing out tracking for XR is nothing, because troubleshooting it can be complex. Rendering frames may take barely any time at all, but making them look smooth in camera as it moves through the space might just take every second you have available before your shoot commences. Aim to make things a known, measured quantity."*
>
> – Vincent Steenhoek

A risk to consider even now (as of early 2022), is that the Unreal plugin is still in a more widely tested and developed state than the Unity plugin is. Some months after we created our plugin, and well after the project finished, disguise's own plugin superseded ours, which is better overall for the wider user community. Both however will need constant, repeated testing, as is the case with any software in development to minimize your risk in production delivery.

Creating the Solution

Our Unity developers worked closely with the development team of disguise to get a working plugin over the stretch of two months.

At the start even the disguise developers didn't know how to hook the two up best, and of course neither did we – it was a collaborative effort to get it there, but we had a functional solution together in time for production!

To deliver a reliable and responsive Virtual Production pipeline, our team needed to take 3D model and texture data from various sources, including the Unity web app and Siemens product departments, and to collate and optimize them for use into a dynamic hosting environment where sales agents could interact with the products.

Our 3D specialists work in InstaLOD for the purposes of optimizing the 3D models, transforming them from high poly models to game-ready assets, which run performantly on the media server's GPUs, all the while keeping in mind the requirements of the client in regards to textures, colors, and animations and being able to control those from the disguise timeline. Figure 14.4 shows our backstage working environment.

> *"As the Unity app (third-party agency developed) for the virtual showroom was still in development while we were already in shoot, we had to prepare each upcoming scene in the shoot on the night before. For this we would get the latest version from a GitHub repository – make changes to the code, to make the scenes compatible with the XR environment. As the Unity app was prepared for navigation with mouse and keyboard, all the user triggered proximity events had to be taken out, so functions of machines and screens within the scene are independent of positions of our virtual camera in the space. We would also get new versions of assets directly from the developing agency, right before the shoot and would have to exchange those assets for a particular scene – quite untested(!). This meant we would be forced to make small changes to the scene on the fly, but because of the real-time environment we were in, we were always able to respond."*
>
> – Urs Nyffenegger

Visit rtv-book.com/chapter14 for the video of the Siemens XR Stage in action

In total the team compiled 11 virtual trade booths and a virtual hosting environment in a stretch of less than three weeks. See Figure 14.5 and Figure 14.6 for examples of the stage in use.

The End Result

All this effort was not for nothing. The results in camera were excellent and the Siemens team and other parties involved

Figure 14.4 Controls behind the scenes at the XR Stage for Siemens. Photo Credit: Chema Menendez.

Figure 14.5 Final composite of the XR system. Photo Credit: Chema Menendez.

were all very pleased with the results. For us as a studio it was an important milestone to achieve, and also to prove to ourselves, irrevocably, that this way of working *is* the right approach for large-scale manufacturers dealing with product presentations and having to address remote audiences. You can choose to build it if it didn't exist before. Just bring enough coffee for the long nights and stressful moments along the way.

Written by the team at Evoke Studios

Project Credits
Evoke Studios
https://evokestudios.io/

Evoke Studios is a leading European studio founded in November 2019 by four like-minded partners: Vincent Steenhoek, Kristaps Liseks, Chema Menendez, Urs Nyffenegger

Coalescing bleeding edge design sensibilities, visual technologies, and a strong sense for the artistic, we make sure

Figure 14.6 Siemens XR Stage with supporting gear. Photo Credit: Chema Menendez.

we deliver for our clients. Leveraging the latest in performance design technologies, taking control over workflow, and connecting the bits in-between is what we do on a daily basis. We deliver the kind of visual chemistry that allows for the realization of any scale production, to be delivered reliably and with insight. It's something we're deeply passionate about.

Project name: Siemens Hannover Messe Digital 2021
Project location: XR Studio cueXR, Bavaria, Germany
Year: 2021

Project team and roles:
Managing Director: Vincent Steenhoek
Technical Director : Kristaps Liseks
Head of Creative Workflows : Chema Menendez
Head of Content : Urs Nyffenegger

Client: Siemens

CHAPTER 15 Production Planning

Every project is unique. Each project team has a mix of personalities and combination of skills. Clients all have their own ways of explaining project expectations. The puzzle pieces of teams, clients, and project requirements change even when projects repeat annually.

Combine these factors with the omnipresent expectation across the entertainment industry that everyone knows the gig and what to do. In an established field, someone starting out will work with experienced professionals to "learn the ropes." Even then, it can be quite intimidating to ask questions about how an established production job is done. Imagine how confusing it is to learn your responsibilities in a field where there are no ropes.

A production depends on the clear communication of strategy and expectations to each team. Effective teams balance simple, detailed communication and taking their own initiative on meeting department expectations. Real-time content creators are only now learning how to achieve this balance. As we learn and establish best practices, we have to communicate more effectively and more often.

Good project communication is best facilitated by what you can control and deliver consistently: documentation. When your documentation is detailed and logical, with clear navigation, it exists as a map of how your team intends to meet expectations. This documentation exists to show other teams how to interact with you to meet their own production goals. It covers everything from system diagrams to content specifications, power needs to headset locations, schedule to contact phone numbers.

We'll start with the initiation of a project with the onboarding of a real-time content team. You may sit somewhere within that team, two or three layers removed from the team leadership. Insight into pressures and responsibilities from the top is important in prioritizing how you perceive the pressures and responsibilities you might face in your particular task.

Client Planning

Depending on the type of work you do, a client will reach out with one of two prospects: we want to build *"some"* thing or we want to build *"this"* thing. Let's start with the latter, where we can expect there is some creative direction involved that defines the "this" to build. Likely, there are drawings to review for discussion and planning. Your team will not start from scratch creatively, even if all you get is a sketch.

From here, we can establish some basic structure of what the client wants you to do: create content, produce a media operations team, organize both? Expectations need to be outlined and agreed to. This is your project scope.

When the scope is clearly defined, a project manager may call and say, "I need a Screens Producer for a two day digital arts festival. All the content will be curated by our Creative Director, including a branding package. I can send over renderings for your review."

Already we are off to a clear start. The project manager understands the services they want you to provide and how to help you understand the demand of the project by providing drawings and alerting you to the fact that video is sourced from others.

You might also get a call that sounds like this: "I got your name from someone at my last gig. She said you can handle screens for RandomFestivalName. It's on this date. What's your rate?"

These types of calls can be intimidating. It may feel as if the caller is implying there is enough information provided that

DOI: 10.4324/9781003206491-18

should allow you to answer the question about rate. There is always time for questions back before you respond. Your job is to turn "some" thing, into "this" thing that can then be made into actionable tasks.

First, who is calling? Is this person a show producer or project manager or a department head for a video team? If it's someone leading a video team, I would focus on understanding what skills they need. That will confirm they understand the tasks they expect you to complete and will outline the rest of the process for you as part of working together. However, if you are getting a call like this from a project manager or producer, consider asking more questions until you are clear on the job demands.

1. When is the project? – If you are not available, there's no point in discussion much further unless you intend to learn enough about the project to recommend another person or team. Find out the project location as this could add travel days.
2. How much time is expected onsite? – This is a good way to find any red flags about the production. While the answer might be a point of negotiation, you might also be told they only need you for one day. Unrealistically short schedules for real-time content projects can point to larger issues.
3. How much pre-production time is planned? – Again, this may be a negotiable point that you need to define, but answers like, "we're going to build it all onsite" are red flags worth noting.
4. Are there renderings or storyboards of the project design available for review? – You can learn a good deal from even the most basic of drawings.
5. What is the expectation of you or your team? – This type of question will often inform how much the client understands about or has experience with video screens and real-time content production. You might receive a clear outline of responsibilities or you might be told to "handle the screens."

Clarifying the expectations of your role can be challenging, as challenging as the role itself. Satisfying an understanding of job expectations will come from trial and error. The more responsibility you have and the larger the team, it's best to define this in a detailed outline of your understanding of the project scope with your budget proposal. Even as an individual on a larger video team, it is good practice to submit your day rate as a deal memo with a written statement about your project role and responsibilities. If you are not at the point in your career that you are responding with budgets or deal memos, respond with an email outlining your fee and your understanding of the project requirements of your time in exchange for that fee. Not doing so could support assumptions that lead to a healthy dose of life experience and non-billable hours.

Budget

The excitement of landing an interesting project or work with a great team is typically followed by the mundane reality of establishing what it will cost. We are in the business of creating entertainment and our livelihood depends on someone paying for the skills we have to make a project successful within a budget.

Besides finding the right team members with the best skills for a job, each team member must establish or accept project fees. Each team member has a day rate. That team has gear costs and project management fees. The production has marketing and talent expenses, venue costs, and insurance coverage. All of this has to fit a budget without running into unplanned costs.

When you are starting in this industry, you will need to establish a day rate that best represents your skills and experience. That day rate has to fit within the larger budget and being told no to your fee request is not personal, even if it feels that way. Talk with your peers and understand what the going range of rate will be in the market before you get on the phone with a potential client. It's also fair to ask if that client already had a rate in mind.

When budgeting for a team, keep in mind there are many costs to account for, not only the expense of paying team member rates. The fees must cover your time organizing and managing the team, but other fixed costs like insurance or project specific rental gear might need to be accounted for as well.

Budgets should also outline how to manage overtime. Every facet of our industry has different approaches to overtime and what a base day rate covers. State laws and unions may be a factor to consider as well. You must also understand if you are an independent contractor or an employee. Some states have cracked down on independent contractors and you may find you want to set up a business entity for your work.

This can quickly get complex and I recommend having a business attorney assist you. An attorney can advise on a business structure and assist with contracts. A legal professional should review all your client contracts as well as your work agreements with subcontractors.

Finally, a budget proposal is a great place to restate your scope of work. This is not to be inflexible in the work you will do, but to reiterate your understanding of what the client expects of you.

Schedule Planning

Establishing the budget will inevitably lead to discussion of schedule as you need to know the number of days onsite, the number of days your team will need to prep, and if your team will be involved with any post production. Each step of this discussion will inform your understanding of the project and the requirements.

Pre-production time, or prep time, can be complex to define. This time needs to cover meetings, documentation and workflow creation, content creation, and other development time in anticipation of onsite work. There is some misunderstanding that real-time content tools mean less prep time. As industry awareness of this technology grows and use of these tools becomes more common, producers are better versed on the prep time requirements. Real-time content preparation time mimics standard content production preparation time.

Onsite, the industry is still learning what the time requirements are for real-time content dependent productions. The platforms used are complex, and systems are interconnected with camera tracking and networked servers. There are many potential data failure points to manage. Any disruptions can eat away at a production day schedule.

Production time onsite is expensive and real-time production teams are full of expert skills at high day rates. When things don't work, it can be costly and stressful as the hours tick by. I recommend including contingency hours in the schedule to account for any new gear, software, or untested workflows. Better still, put pre-production time in the schedule to test new workflows and avoid expensive time during production.

Review

Things change as a project moves forward. As part of a good communication strategy, it is good business practice to check in with a team leader and/or client regularly. This will give you insight into any changes that might affect budget or schedule, or that the production is progressing as planned. Silence from your client or project supervisor can mean the show is on schedule, or it can mean they are waiting for you to complete a task you didn't know was yours to complete. The occasional "oh, by the way" from your client or team leader is common. Expect it now to avoid the stress of surprise later by having contingency time and budget set aside.

You will need to anticipate and account for changes in scope driven by factors outside your control. Perhaps a performer backs out of the show. Maybe a sponsor has a critical demand requiring new video content. These changes fall outside the agreed upon scope and will increase team work hours. When that happens, discuss the change in fee or expected overtime with the client as soon as possible.

Occasionally, there is a change in understanding of the existing scope. This happens most often due to a misunderstanding or misinterpretation of the scope requirements by the client or by the supplier. Each situation has to be reviewed for its effects to the project and how to best find a resolution. We want to keep our clients happy, and make allowances within reason.

Workflow Outline

Now that we have outlined the budget and schedule, the next step is to review the process ahead to complete the project requirements. What are the steps necessary to prepare you, your team, and the rest of the production when it comes to working with real-time content? While there is no fixed approach, there are basic tasks to tailor to each project and how real-time content will be used. Is real-time content used as a part of the content creation toolset or is real-time content critical to the entire production strategy? This section will highlight tasks that cover all types of screen environments for a range of real-time content uses.

Scenic and Storyboard Review

Some review of renderings and technical drawings has occurred in the budget and schedule planning. Detailed review is recommended as your partner design teams finalize project elements. Scenic drawings and storyboards eventually become technical drawings and design specifications for the screens. The physical infrastructure of the screen system affects what you can and must do, from design to engineering to operations.

These drawings will define the number of video signals it takes to cover the video surfaces, how many pixels are necessary for good image quality, how complex the content design might be, and so on. Understanding these points will inform choices from the type of real-time content creation software and to the hardware platforms used to process content.

Review renderings and design documents, and make note of questions you have for the teams you work with. Information important to the work you do can come from the set designer/

art director, the technical manager and screens engineering team, and the team handling any real-time data inputs like camera (telemetry or signals), audio, skeletal tracking, IR tracking, or user inputs from other computer sources.

Your role might focus on a narrow set of information or be affected by every aspect of the answers received. If you are missing information to complete your specific tasks, communicate the impacts of that missing information to those who need to know. Only you have clarity into what is keeping you from completing a task. Don't expect your challenges to be obvious to others or what insights you need to eliminate an issue.

Creative Discussions

Drawings provide a variety of technical and creative insights. Concurrent with technical planning, design discussion will begin with your creative partners. You may be supporting designers with specifications and delivery guidance or providing design direction to content creators. In either case, creative discussions should start at the top with the producers and creative directors of a project.

Creative discussion is typically supported by a series of design development progress documents that show the creative direction of the design team. Once approved by the event stakeholders, these documents get passed on to supporting team members to become build documents and/or act as the design guides to create digital assets.

If your role is to build these digital assets, you may need to sketch out your understanding of the design direction and share the file development for review. This can come as creative stills from a work-in-process (WIP) project file or full pre-visualizations. I find 3D previsualization essential to good feedback. Content presented as it will be experienced on a stage or through the camera lens is a powerful approval presentation tool.

If the format for WIP approvals doesn't fully inform the project stakeholder, the potential exists for these approvals to come from an incomplete understanding of how the creative will be used. For example, a rendering of AR content as a still might look beautiful, but if it fails to show how that AR content interacts with the camera or scene, you may not convey the full creative intent. A producer might approve the file and later react onsite by saying, "I didn't know it was going to look like that." It is unlikely they will feel any responsibility for the time it will take you to correct the issues now visible to them. Find a meaningful way to have informative, creative discussions.

The power of many real-time content creation tools is they also provide a perfect environment for previsualization. Many media server platforms do as well. As more production partners have become versed in 3D skills and language, it is now easier to share 3D assets for review, through standalone files or via a web link. These tools will keep advancing for use by clients who have limited knowledge of how to use 3D, but will benefit from a full 3D previsualization. Whatever route you find, good previz practice will lead to better use of your time onsite and limit creative miscommunications.

Team Resources

You have a budget, schedule, and tasks. How are these allocated? Who ensures that goals are met on schedule? Who in your team are you telling when goals have been met?

Depending on the size of the department you work in and the responsibilities for your team, some structure is necessary to move forward strategically to meet team goals. For example, if the system programmer is ready but content files are not due to be delivered until the next day, that's a poor use of the programmer's time and team budget. Similarly, if content is ready early, but no one knows, the team can't make use of this time savings.

It is up to the team producer or project manager to outline target goals for the team relative to the production schedule and expectations of the video department. This should also include a state of readiness for rehearsals and client reviews. These internal targets will establish a fine-tuned schedule for the team.

Build and Publish Content Production Workflow

Within the video department, there will exist an outline for content production and delivery. Planning and describing a content production workflow is a very particular and complex task that is essential to good inter-team communication and efficient content delivery.

> If you are interested in learning more about building a content production workflow, check out my Screens Producing and Media Operations textbook. More details are available at rtv-book.com/chapter15

The Content Production Workflow for real-time assets will outline essential technical information for the content production team, including image quality requirements, needed control variables and interactive inputs, and polygon targets for 3D assets. In traditional content delivery, important technical concerns are raster sizing and pixel density, codec and frame rate, as well as color space. File naming, revisioning, and delivery methods will be covered in both cases.

The content production workflow needs to be simple to digest for the user, easy to navigate and clear, really clear. The work is complex enough; we do not need to create more complexity in the work's description.

Development and Testing, Review and Reiterate

At this point, the team has direction, schedule targets, and a process to get there. Now it's time to build something. With real-time content creation, this will involve building out content files, reviewing them with your team leaders and clients, receiving new direction or refinements and repeating the process until the work is approved. When content depends on real-time inputs, the approvals process will include data input simulation in previsualization or testing with real data.

The familiarity of the team or challenge of the task will determine the time needed for this development cycle. Building something entirely new is exciting, but it is essential to get as much real world testing as possible before use, especially if the use case could result in a studio full of top paid professionals burning hours while a bug gets worked out.

There are several ways to simulate testing environments when your work depends on complex or expensive input information, like cameras. We've seen examples in the case studies that use simulated data to stress test a real-time content platform. If real world testing is the difference between success or failure, then this should be part of the budget plan and the client should be clear about the risks of limiting test time.

We work in a field where we are constantly pushing the boundaries of our experience and capabilities. We also enjoy a working environment that is based on collaboration with other creative and technical thinkers. In our next case study, we'll see the challenges and strategy dandelion & burdock develop as they work on an interactive installation.

Case Study – dandelion & burdock – Johnnie Walker Princes Street Interactive World Map

Overview

In spring 2021, our team was approached by BRC Imagination Arts (BRC), a global strategy, design, and production company, to create animation and real-time visuals for Johnnie Walker Princes Street, a whisky visitor experience in Scotland. At the site, a beautifully refurbished former department store, audiences were to be led through a series of educational and experiential rooms in groups of approximately 20 people. These guided tours commenced at timed intervals, controlling and directing footfall. Accessible to any visitor, the venue's atrium featured a light and video art installation, vertically spanning across multiple stories.

dandelion + burdock predominantly created content for the atrium piece (see Figure 15.1). The task entailed integration of video and generative sources, programmable virtual lights, and customizable looks for special events. The real-time and interaction platform was Notch. All hardware and essential software across the site had been specified well in advance.

Additionally, we created an interactive world map on a wide aspect LED screen in a waiting area of the venue. Initially, the map task had a lower priority since it appeared to be easier as it had clear creative direction, required less interdepartmental alignment, and had a less challenging workflow.

This article will focus on the runner-up, as it allows valuable insights to working on real-time projects. Perhaps more so than the large centerpiece.

Brief

Before a guided tour begins, visitors gather in a welcome area. The main screen in this room displays a flat world map, scrolling from right to left, changing randomly through four color themes along with the ambient light in the room. Countries in the central area of the screen will randomly highlight and reveal respective salutations.

Awaiting clearance from the adjacent tour room, a host entertains the audience by asking for the visitors' home countries. On a tablet the host selects a match. The map now completes the current greeting animation, focuses on the selection, stops scrolling, changes to a discreet color preset, the ambient light transitions accordingly, while the map zooms into the matching country, which is becoming a stencil alpha hole for a brand video clip playing underneath.

The clip has a custom length based on the brand's history in the region. Operator inputs are suspended for the duration of the video. At the end of the clip, the system dissolves the video's tail into a zoom out on the map. The world map returns to the idle state and is ready for a new input.

Figure 15.1 BRC x Johnnie Walker Princes Street-93 – Around the World. Visitor experience created, developed and produced by BRC Imagination Arts, brcweb.com.

Some assumptions made for the functional description are:

- The color schemes, highlights, transition speeds, greetings, and selectable entries can be adjusted onsite, ideally without specialist programmers.
- The custom brand videos are hosted in a media server system, while a Notch block is rendering the map, calls the respective clip and handles the transition into and out of it.
- The block communicates to the lighting system, not vice versa.

Preparation

With clear objectives and specifications, the map task seemed to be well defined, although alternative technologies might have been more convenient. After receiving creative assets we estimated completion to require just under 20 days for a maximum of two artists, following the client's suggestion that the map was a straightforward recreation of a set of provided motion samples.

As a foundation for the welcome-interaction, we set up an online list of countries and regions, linked to an equirectangular master vector map.

In the list, the client completed metadata such as color preset selection, greetings, active components and video durations, in and out points. Later on, this would also serve the onsite team as guide for their integration.

The vector maps gave us flexibility for styling, resolution independence, and mapping type transfers. As a sandbox environment, we had an OSC server to simulate interactions and visualize outgoing signals from Notch.

It was helpful to define all variables we would either want to receive or send and consider the modes the system could be in: *off, on, idle, transition-in, playback, transition-out*, and link these to tasks.

Process

Despite the necessary preparations, our first mockups came about quickly. Initially, we exported countries as a texture sequence, driving a highlight selection by frame number. Map scrolling was achieved by texture offset. Greetings required parsing a local copy of the master list and timing a font animation.

Problems arose when putting elements together for the first time. For example, the 2D texture offset would not reliably connect a greeting to the required position. The texture selection forced us to use bitmaps, which could not be focused on without eventually exposing pixelation. We realized that each country's zoom in point would need customization, and that the vertical layout of each greeting required layout rules. Finally, the inhouse font posed great limitations. Latin and cyrillic character sets were available, but at a minimum we needed simplified Chinese, Japanese, Hebrew, and extended western characters and ligatures to display the most important greetings in their respective writing system. The estimate ranged from about a few dozen to a thousand missing characters.

Working on parts of the project, achieving success or identifying roadblocks was slow. Our attitude had to change to move the project forward. It was clear now we had underestimated the time needed to complete the creative brief.

We decided that each part of the brief was a negotiable module. A module would not be either solved or impossible, but considered a fixed or soft requirement. Our success criteria would be the creative intention, not a verbatim implementation. Fixed requirements needed to be met at some point before delivery, soft ones could be exchanged for alternatives or removed. As a result, each part of the brief and mockup animations had to get scrutinized for hidden or previously unidentified assumptions.

To solve the accurate positioning and pixelation, we returned to the vector file and built a 3D scene from it. Each country got turned into a polygon and was imported into Notch using geometry enabled parenting objects, like the greetings module, or a focus-null object. Outlines now stayed perfectly sharp at any zoom level. Our diversion meant a lot of extra preparation, but offered benefits in turn. In Notch however, the node tree got extremely large (see Figure 15.2, Figure 15.3, and Figure 15.4), which made navigation and teamwork harder. Remote collaboration required diligent annotations, coloring node tree regions, and frequent handovers.

> Visit rtv-book.com/chapter15 for full size images of the Notch node trees used on this project

For the font problem, we did not find any wildcard solutions. Single fonts could not hold all characters, but building fallbacks

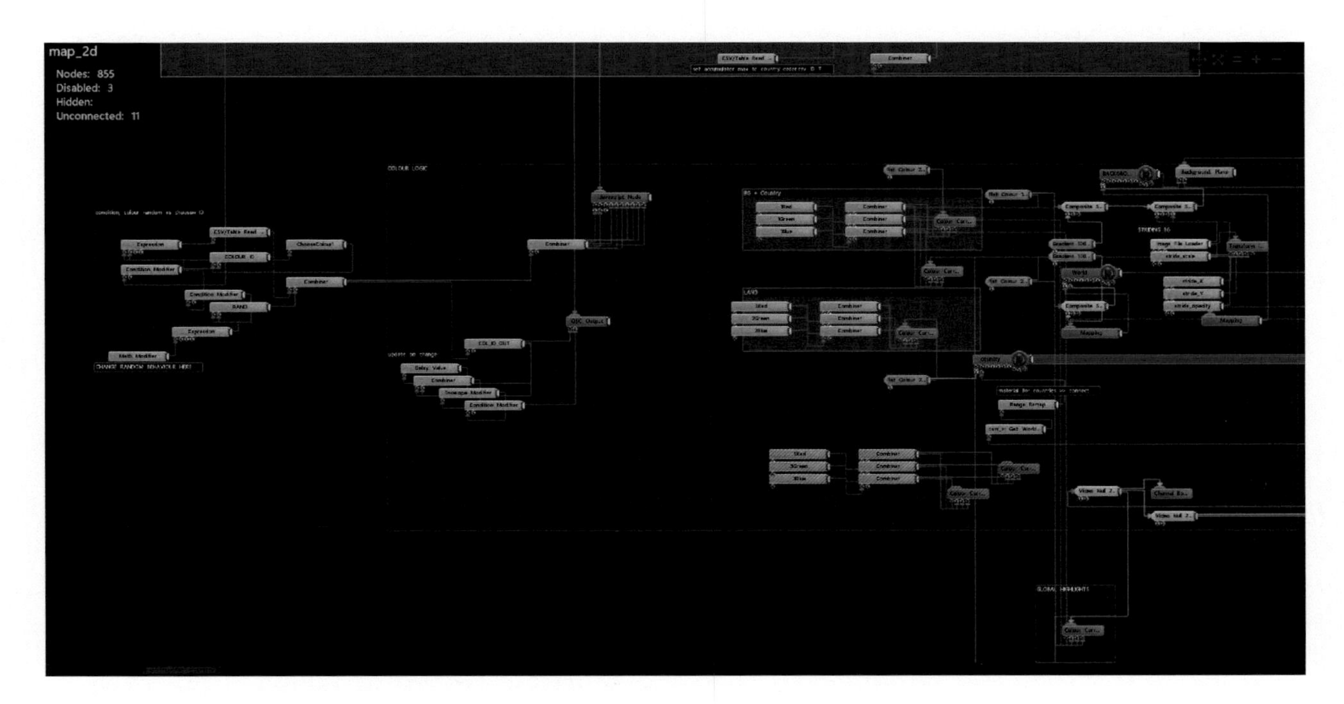

Figure 15.2 Detailed view of random color function (left) and shader setups for background, foreground, and highlights states, fully commented by region. Source: Image Credit: Nils Porrmann.

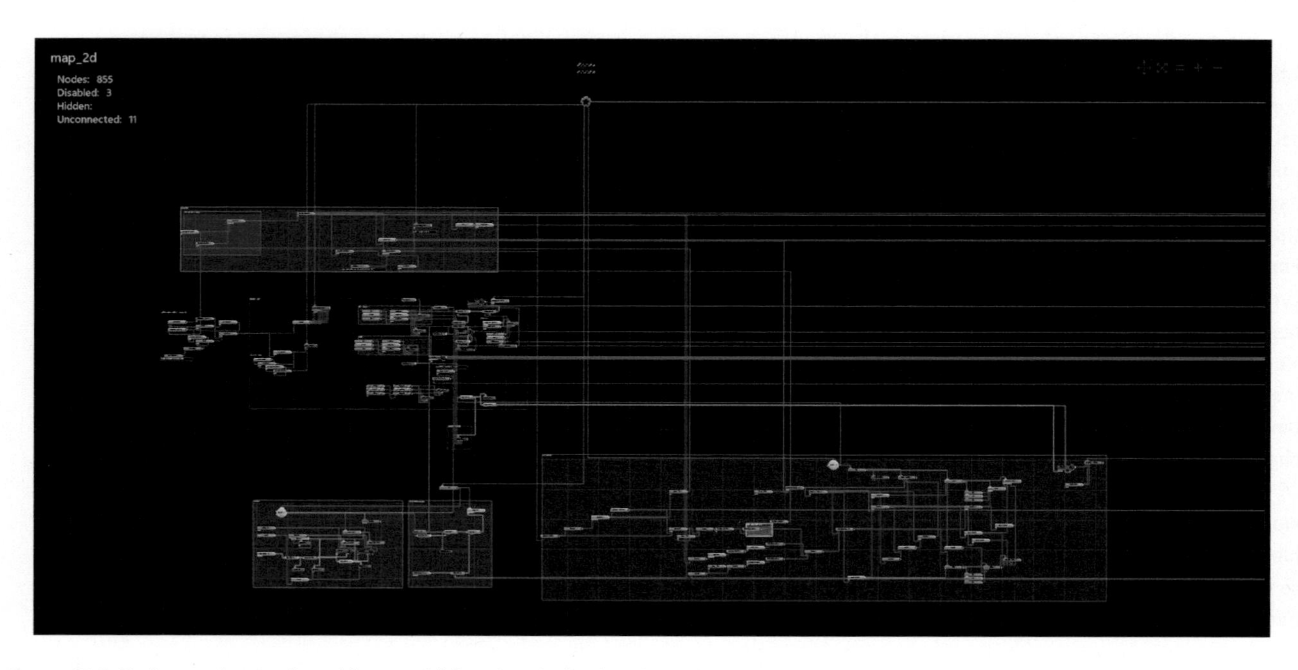

Figure 15.3 Medium node tree view with essential functions (yellow), color and shading (purple), compositing (green) and greeting highlight module (dark gray). Source: Image Credit: Nils Porrmann.

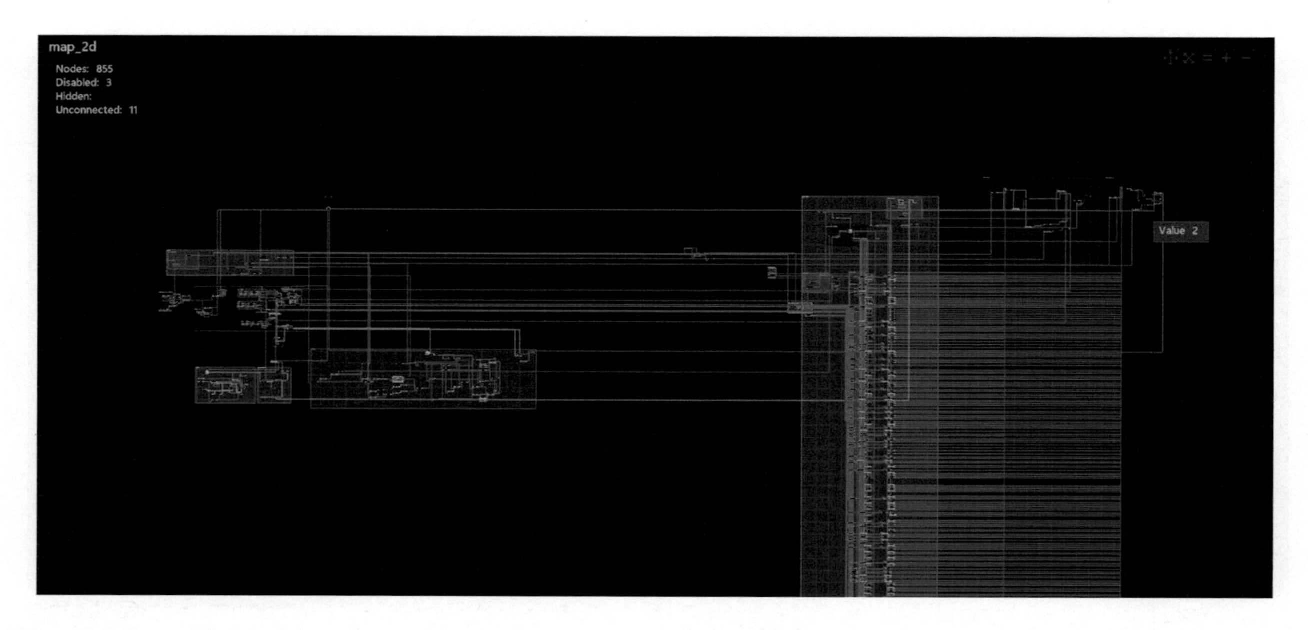

Figure 15.4 Almost full view of the node tree with vertical country and zoom zone blocks (brown). Source: Image Credit: Nils Porrmann.

for font selections was tedious too. We had to approach the BRC team: Our suggestions were:

- Creation of each greeting as a bitmap or vector artwork; a lengthy and non-parametric path.
- Additions of the most common special characters to the existing font, requiring the involvement of typographers.
- Transliteration of all salutations to a western alphabet, needing extra checks and phonetic research, but restoring flexibility.

Looking back at the brief, countries in the center of the screen were supposed to highlight randomly.

This entailed two tricky requirements. An array of random indices, updated with the most recent highlights, and checked against a central hot zone function to limit the upcoming selection to the center of the displayed map. After much experimentation, we found that both options were possible but not viable to develop into our block in the time given.

The work in progress designs revealed that highlights could be timed throughout multiple revolutions of the map. Instead of a truly random function, we now made a list going over all possible entries and curated a pleasing loop. This loop would be interrupted by an operator selection, and reset on completion of a video clip. The curation also resolved center zone logic as elements were chosen by legibility.

A final example to consider was the zooming country revealing any respective media clips. Notch's depth buffer provided the data required, but in the media server our initial paradigm of opening a window onto a lower layer was not providing the quality we expected. Instead, our most reliable solution was porting the media server texture into a Notch video input and then composite the clip on top of the layer stack in the block. Though not a complicated fix, the solution matched the visual intention; it was nonetheless an inversion to the description in the brief.

Delivery + Opinion

Building a map is not sorcery. It is not even a particularly glamorous example, but the issues discussed here are synonymous with a type of problem solving within creative services. The client will approve deliveries that meet their expectations. How our team, you, or myself got there is of secondary interest. The priority is to work with clients to realize their vision, while also recognizing what aspects of that vision can be evolved to make the project successful within the time and budget available. Figure 15.5 shows the same space from Figure 15.1 during installation.

On one hand, I think that motivation is a great starting point for this type of work. The way to see it through is resilience. Success? Move on. Fail? Reconsider the issue, and fail better next time.

Figure 15.5 Onsite full raster test grid and media server integration (Notch in 7th Sense). Photo Credit: Nils Porrmann.

I realize that a common method in most of my work and certainly in all these examples is reframing. Breaking complex tasks into bite-size chunks that can be done, delegated, or dropped. Apart from picking the low hanging fruit first while wondering about and contemplating the next steps, differentiating fixed from soft requirements helps me to maintain momentum.

These practices inevitably lead me to better planning, as I have to account for time and thought to reassemble the chunks and communicate with clients and coworkers.

Some strategies we have explored here are *rebuilding* (3d map), *delegation* (font + greeting), *alternative* (randomness), *redefinition* (alpha stencil), *deletion* (center hot zone). Whatever the name, I find it useful to come up with these categories and plot my tasks against them.

In all of this, it is great to have an understanding client and check feasibility with them along the way. If time, resources or budget become concerns, they need to be communicated. Although the map required almost twice the budgeted time – clearly our loss – a foundation of trust and honest exchange between BRC and dandelion + burdock allowed for collaborative problem solving, mitigating risk for both parties.

By Nils Porrmann, dandelion & burdock

Project Credits
dandelion + burdock
https://dandelion-burdock.com/

D+B create content and consult for ambitious brand campaigns, digital first experiences and live entertainment projects. With a uniquely design driven constellation of departments, D+B actively seek projects that integrate disciplines, creating exceptional technical solutions for compelling concepts in special event scenarios.

Project name: Johnnie Walker Princes Street Interactive World Map
Project location: Edinburgh
Year: 2021

Project team and roles:
Design: BRC
Notch Artist: Harrison Mead
Notch Lead + Onsite: Nils Porrmann
7th Sense: Lucas Harrison + Andy Bates

Client: BRC Imagination Arts

CHAPTER 16
Obstacles to Success

Creative video is a new production paradigm. As a content creator or systems engineer, your role and responsibilities may be well understood. Equally as likely, you may be the mystery video professional keeping an essential computer running behind the scenes. Even if you are part of a content production or support team that has been working in creative video for decades, real-time content creation has upended processes and introduced many new production roles.

Each entertainment design discipline goes through a maturity cycle as it integrates into the larger production process. Before there was a lighting designer, there was a technician who made sure there was light on stage. The art of Lighting Design came later. Video has had similar utilitarian applications on the way to becoming a design practice. As that practice develops, we identify new roles and responsibilities to elevate that practice.

The creative video design discipline is currently experiencing a rapid rate of change and disruption. Practitioners have experienced an incredible challenge trying to build a shared process and common language around this new body of work. As soon as we develop community knowledge for one process, we introduce a new technology that disrupts that process.

Still, we manage to learn the very newest practices from each other and through inspiration. We see examples of the creative and technology in use by other teams. A small, innovative group of professionals share their real-time skills and successful workflows. However, I believe we need to do more than whisper tricks of the trade to a few trusted partners. We need to educate each other and our clients on best practices and engage manufacturers to help clarify what makes us and their products successful.

Knowledge sharing hit a low in the early 2020s and the impact has left us unable to support the current rate of growth in Virtual Production. It's time to return to open sharing and community education or else we limit our potential. This chapter explores ways we can avoid becoming obstacles to our own success. We have covered good communication and team structure when building a real-time content driven project in the last chapters. Let's now look at some challenges facing the community when building a real-time project and developing a team.

Community Education

During the rush to invest in successful VP practices in 2020 and 2021, the open sharing of techniques stalled between users. This was a natural side effect of the limited number of active projects and the investment required to meet existing work demands with newly developing solutions. Early adopters of VP integration techniques put in an incredible number of hours developing successful workflows. These hours were often at their own expense.

Successful workflows guaranteed teams' access to the few projects available. Many teams invested in the development of proprietary solutions to make their Virtual Production process function. Those investments created financial and mental debts. These teams naturally protected their new knowledge base because it was expensive and exhausting to secure. This has led to a lack of available skilled real-time technology talent to meet increased demand as production returns to pre-pandemic levels.

Whether it's the previously depressed job market, the complex skills required, the volatility of a developing production practice, or some other challenge, we have found ourselves short on VP practitioners and real-time content creators. Technical knowledge is only part of the story, as we also need people versed in production culture. Our work is a cross-disciplinary practice, however our entertainment production culture is not for everyone. When we look for new team members with game engine skills, we also need to educate them on the expectations of our working style.

We need to upskill professionals who already have a mature understanding of entertainment production. This includes

DOI: 10.4324/9781003206491-19

making an investment in the next generation of skilled entertainment technology students who want VP careers. The descriptions of projects like those in this book's case studies, engaging with fellow community members through forums and conferences, and inspiring one another by speaking openly about our successes, is part of that process. I'm not encouraging anyone to give away intellectual property, but I am asking for more shared awareness of how this work is done.

Education not only serves the production teams doing this work, but also those who are hiring these content teams. Client education exposes the complexity of what may otherwise look too easy. When everyone involved in a production better understands the requirement of executing real-time technologies, we as a professional community are better supported. Let's examine the client related factors in more detail.

Inversion of Process

Real-time technologies change producers' anticipated production schedules. When using traditional content production, the production schedule allocates time for edits, re-rendering, and re-delivery. These steps are simplified or eliminated in the real-time content workflow, saving considerable time in production. To realize these time savings, real-time content needs pre-production development time to review content pre-visualizations, plan useful control variables, and test the system.

A VP project using ICVFX also shifts budget and schedule allocation. VP eliminates time spent in post-production applying traditional VFX, but introduces a need for additional pre-production time to prepare the content for ICVFX use. When a producer is unfamiliar with VP, you may have to work to explain and outline these differences.

Consider the perspective of those accountable for the production budget. If they have successfully produced many events involving creative video and screens without leveraging real-time solutions, what incentive do they have to change their approach? You must come prepared to explain what makes real-time content production different and how additional prep time ultimately improves the onsite experience and adds value (either faster onsite decisions or improved production values).

Digital Is Perceived as Easy

Across the broad swath of digitally enabled work, we find the odd (to us on the inside) expectation that the computer does the "labor." If you build websites, edit photos, or translate documents, there is a tendency to see the human component as purely supervisory. It can be very challenging to quantify the human side of digital labor. The creative video industry suffers from this same misunderstanding.

Clients routinely underestimate the time and people necessary to complete a real-time content project. Don't take offense when you are at the receiving end of such misapprehensions and misjudgments. Instead, take it as an opportunity to educate the client. This may mean declining a project with unreasonably stressful time and expense expectations.

Your challenge is to make fair time estimates for completing the project as you understand it. No client will readily accept an hourly rate with an undetermined number of hours. Real-time content and integration teams must provide a clear estimate of the labor required. This is best learned through experience. When getting started, senior team members and department heads will establish your time and budget. Take every opportunity to understand the time expectations of your field and understand when certain tasks can be optimized, or benefit from more time.

Another perception challenge for digital labor is that many producers see computers as tools containing infinite possibilities. Click one button and it just works, right? Set good boundaries with your clients and production partners in order to keep unrealistic expectations in check. This means reminding them that a human does the thinking before hitting the keys that "make it just work."

Creative video production teams make a habit of pulling off feats of magic. We are fantastic problem solvers and hard workers. The problem comes when we reinforce the belief that we can do anything. When presented with a request that is time intensive or potentially destructive to the work already in place, don't jump to say yes. Make sure you understand the goal of the request. Share the full impact of that request. You can say "yes" to meet your client's desires and also intelligently push back with sensible boundaries.

Strategies for building useful production boundaries are all rooted in good communication. When faced with a challenge or complex request presented by your client, take a moment to consider the impacts to the work in progress.

- Start with your own comprehension.
  - Did you as the practitioner understand the request correctly?
  - Maybe it's simpler than you heard. Take a moment to fully digest what was asked before answering.

- Consider alternative approaches.
  - Is there an alternative that is less time intensive and will achieve a similar result?
  - Offer a variation. Lay out your understanding of the initial request and its implication against another approach.
- Clarify the consequences of taking action on the request.
  - Is the request outside the capabilities of the current system design?
  - Be honest about the tradeoffs. Will you need overtime to make it happen?
  - Be clear about what it takes to say yes. Is your alternate idea a little less perfect but a lot more affordable?

We all want to say yes and meet the expectations of our clients, but to do so at the expense of the stability of the entire project has to be a considered factor. Working extra hours or adding gear can solve a challenge, but adding too much more of anything too late in the process risks the overall stability of the project. Make sensible and truly supportive choices.

Real-Time Can Be Time Consuming

Among the greatest real-time content production myths is the belief that real-time content is fast to produce. Real-time content is fast at one thing: rendering.

Something that renders 'right now" can misrepresent the complexity of the rendering process. The information and components involved in making something render in real time can be immensely sophisticated. Well-built complexity looks simple and takes time.

Unfortunately, in order to make time requirements clear to our clients and project stakeholders you must demystify our magic. Be transparent about the human effort and the extensive tooling that make real-time a reality. Expose the planning and structure required to make real-time function reliably.

However, demystifying our work does not mean restricting the creative process. Creative video can't be infinitely adaptable, but it can be very flexible with good planning. Communicate with your clients to understand the difference between fixed project requirements and areas with room for change. Real-time needs the same creative development time needed by all creative video production.

Team Growth and Wellness

The last obstacle to success when working on real-time projects is, well, success.

Success usually means long hours under high pressure doing complex tasks. This work can be exhilarating when you figure out a complex solution that wins the day and devastating when you have no idea why the trick you invented last week refuses to work this week. Production work can be an emotional roller coaster ride of small wins, big frustrations, great expectations, moving goal lines, and amazing spectacles.

Not everyone enjoys every type of video production work. If you find building real-time creative video for concert tours unpalatable, maybe you will find work on interactive museum installations more to your liking. There are many options and production environments for these skills. When you find the place that most inspires you, remember you still need to rest and recharge.

Knowing when to take a break is one of the hardest lessons of inspiring work. You will discover great comradery when you work hard to overcome impossible challenges. You won't want to disappoint your friends by not being there to help. This can lead to saying yes to every job that comes your way. And too much work is a recipe for burnout. Please remember your friends need you to be healthy so that you can support them as often as you can.

Take breaks. Schedule your down time and notify regular clients so they know when you might be unavailable. Make room for daily non-work activities to keep yourself healthy and inspired, like meditation, exercise, reading, art, community, or family. Burnout makes going to work painful, removing the joy from work that once made you passionate. Take care of yourself.

Another aspect of self-care is making room for your technical and creative growth in your work. Explore new software or create a new workflow for a tool you already know. Self-motivated professional development is immensely satisfying. I've gone to a company headquarters for specialized training to improve my understanding of technology, one that I might not even use! If an experience helps me understand the work I do and the team I work with, it is time well spent.

Let's read about the team from SAVAGES and how they use their downtime to improve their practice.

Case Study – Savages – Practice Makes Process

A Case for Intentional Practice

My studio, Savages, started out as a creative execution firm serving real-time content needs. Creative directors and producers would pass us a vision via mood images and time-stamped storyboards. In turn, we build the imagined worlds and effects.

It is an exciting seat to sit in for several reasons:

1. Media technology is evolving at a rapid speed.
2. Content creation software has grown diverse, and each program offers extensive functionality.
3. Projects have variable timelines, quality requirements, and production scales.

These considerations, when combined, make each project feel unique. Unique is often exciting and rarely efficient. However, for most real-time projects, efficiency is the key to success. We see this juxtaposition to be a prime challenge for digital creators.

Since the projects are unique, it is up to us as technical creatives to make our processes efficient and our abilities dependable. We need to work with our tools to know what we can build and how long it may take. It is not enough to hammer a nail into a block of wood. A carpenter gets better at making chairs by making chairs. Similarly, we organize weekly work-study sessions around internal curiosities and creative concepts. Because the team controls the timeline, we can zoom in on any area that needs refinement.

Intentional practice, for us, begins with setting a goal. Given our focus on integrated virtual environments, we begin with a setting or a pipeline challenge. Next, we discuss how we might approach the project and create a visual document to track our plans. Then, we get to work. Afterward, we discuss the process and results. The goal of this, of course, is not whatever we have managed to put on the screen. The real goal is individual growth and team efficiency. So, pausing to clarify what was an easy reach or a difficult stretch helps create benchmarks for future reference.

For example, this winter, we decided to re-approach an old project brief calling for fireworks created in Unreal Engine's particle generator – Niagara. Figure 16.1 and Figure 16.2 are example stills of the fireworks content.

The first time around, we added the Niagara firework system to a blueprint and set its spawn rate from zero to one via a level sequence timeline for real-time playback. The process served the project's needs, but the build was cumbersome, and the playback was not very flexible. So, this time, we wanted to do more than trigger a set level sequence with looping fireworks. Figure 16.3 describes the approach.

Figure 16.1 Niagara generated fireworks from Unreal Engine. Particle System build by SAVAGES.

Figure 16.2 Niagara generated fireworks from Unreal Engine. Particle System build by SAVAGES.

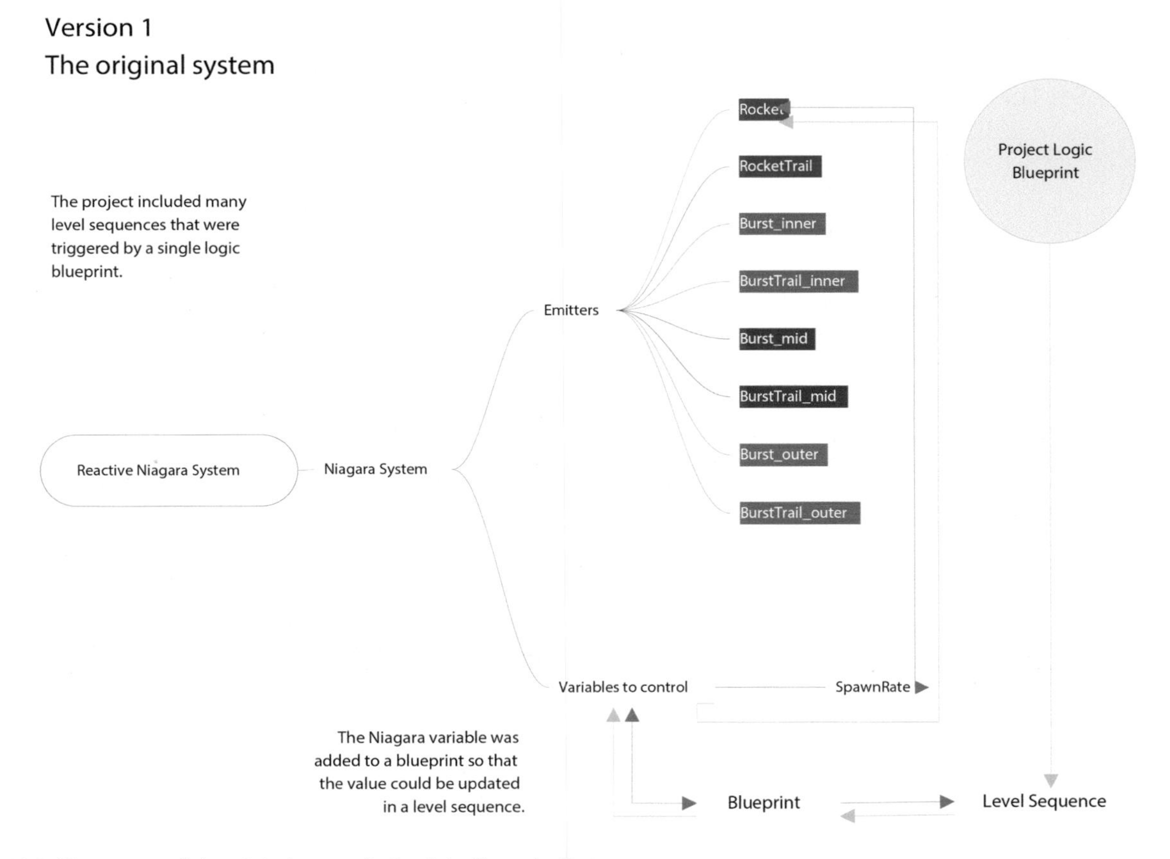

Figure 16.3 Niagara map of the original approach. Graph by Alexandra Hartman.

We wanted to give the effects some *life*. Our goal – beyond stretching technical abilities – was to experience three different ways of controlling a Niagara firework system: an external control, an internal clock, and active user input. Because we wanted to compare the utility of each playback type, we would also create a menu to make it easy to switch between the playback types. Finally, we wanted to clean up the process and create a project tutorial for later reference.

We prefer to create flexible systems. That generally means working parametrically. For this project, we would build independent blueprints and use various communication styles to pass our variable values.

The original project contained five fireworks, each uniquely styled. While that scene met the 30fps requirement goal, it would have run better with fewer systems in the scene. So, we decided to connect the Niagara firework system to its control blueprint with structures and data tables. Though this required us to add several additional steps, we suspected the change would benefit us in several ways. See Figure 16.4 for the first step in the new approach.

The original system came together in three days. We decided to give ourselves the same period to accomplish the new, more complicated build.

First, we built the firework system. Then, we created the variables we had determined necessary. After plugging those variables into our system, we made two structures. The first structure carried the variables' names. The second carried the variables' types. In Unreal Engine, it is possible to create a data table of presets driven by a structure. The structure essentially defines what type of variables will be in the data table. So, we made a data table of values based on the structure of the variables' types. Next, we needed a blueprint to control the Niagara firework system and update it with variables read from the data table. In the construction script, we brought in the variables' name structure and the data table of style presets. Then, we set both to correspond to an enumerated list.

Since we made styling-ease a goal, we added a "custom" row to the enumerated selection options in addition to rows that represented the data table. That addition enabled us to have one flexible option for previewing new parameter values before adjusting the data table.

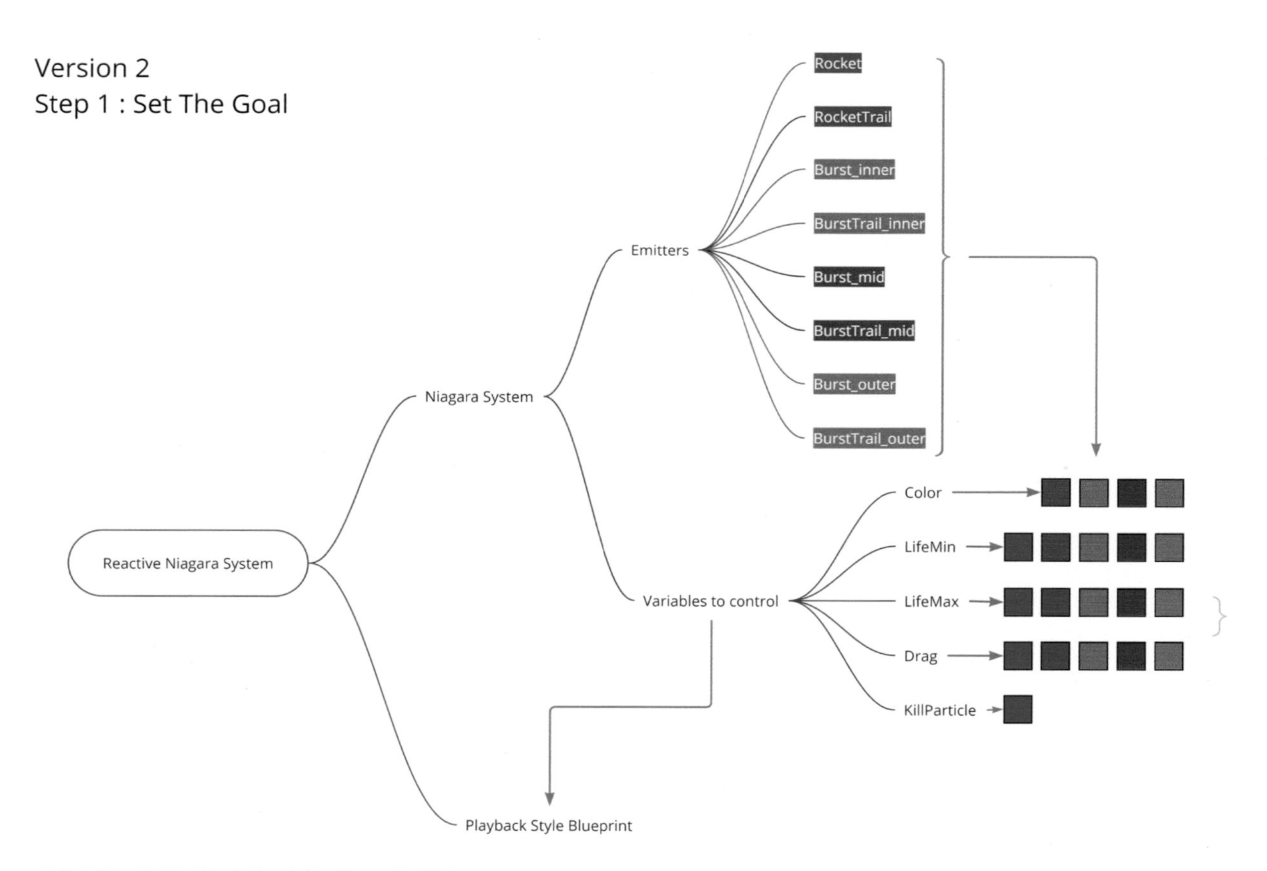

Figure 16.4 Step 1: The Goal. Graph by Alexandra Hartman.

Next, in the event graph of the blueprint, we built a simple switching system. This system would select a random style from the data table and set it to the Niagara system. Then, it would communicate to the fireworks when to launch or when to explode.

In separate blueprints, we built an OSC receiver and a timer that sent a message every set number of seconds. Both connected to the Niagara control blueprint with a blueprint interface.

Finally, we added more functions in the blueprint interface to connect the menu widget to both the OSC receiver and the timer blueprints. Essentially, when the user selected a playback style, the other options would be blocked from playing. See Figure 16.5 for the plan.

After a practice project is "complete," we look back at the original goal and access our process. What was easy? What was a challenge? Did any new questions come up? Did we learn anything about the balance between style goals and technical decisions? The list of questions goes on, but the point is that setting aside time to process what we have done is really where the value of practicing is most noticeable.

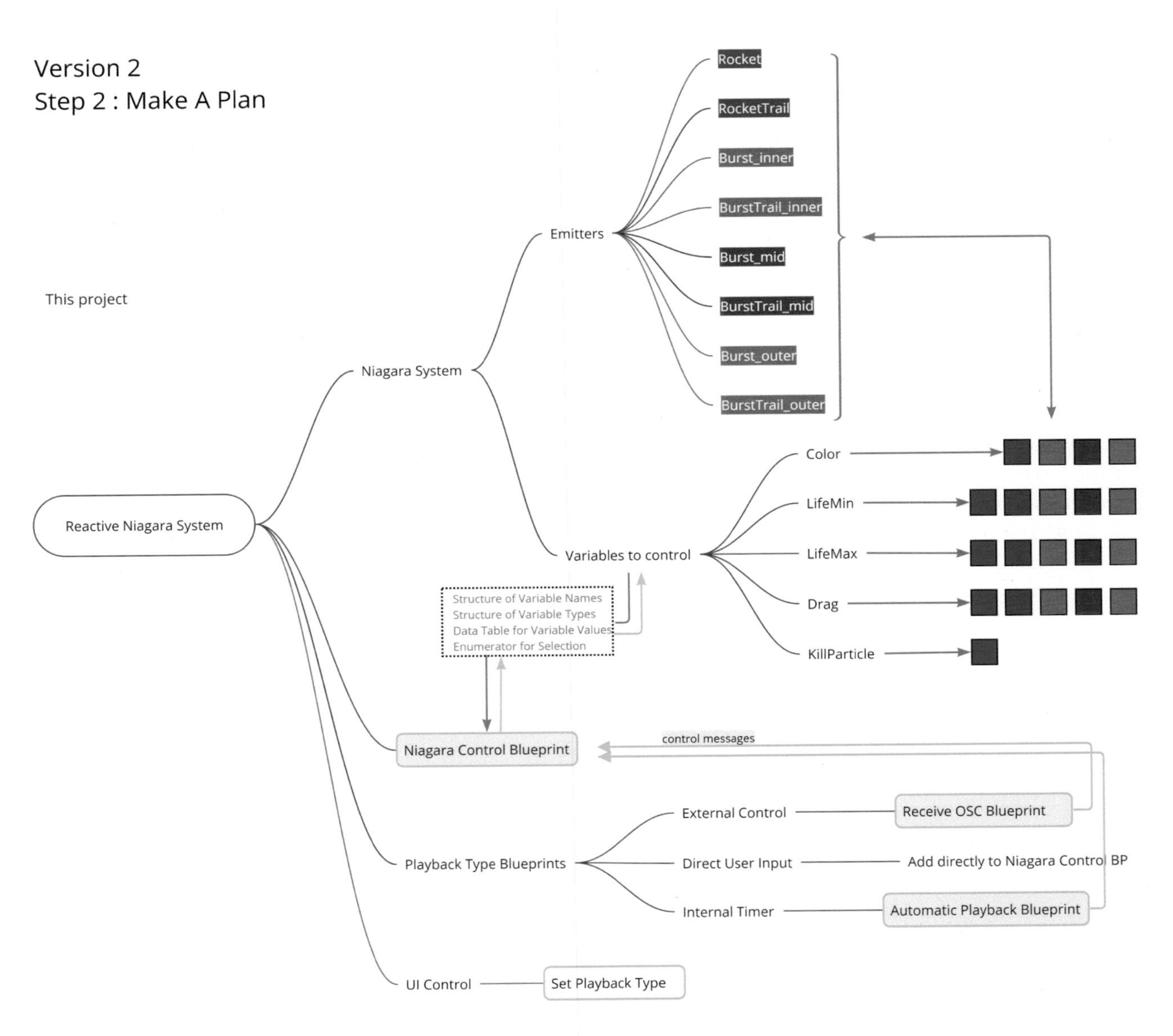

Figure 16.5 Step 2: Make a plan. Graph by Alexandra Hartman.

In the former example, our goal was to drive the Niagara particle system from blueprints and observe the stylistic effects of various technical processes.

Some notes on the findings:

1. Data-table-driven variables increased overall efficiency.
2. A randomly switching system wanted five or more style options in the data table for variety.
3. Audio-reactive OSC messages provided very active and irregular triggers, which effectively gave the Niagara system a lot of life.
4. A timer-based solution felt rigid with a set decrement value, and it improved by using a random float in a range.
5. The user-input playback was dynamic, but the firework styles did not switch in a way that gave a multi-color effect as the OSC and timer-based options.
6. Adding a light renderer to the fireworks did affect the frame rate, but the effect was worth the compromise.

There were also a couple of points we found for improvement. First, we had not enabled our system to stop or start fully. It was not explicitly part of our goal. However, in playing with the system, we wanted the functionality. We decided to quickly add a button to change the visibility of the fireworks. Changing visibility was not a perfect solution because the fireworks disappeared and reappeared – the particles did not fade naturally. Solving this required an additional boolean variable in the Niagara system.

The firework project example came together more quickly than expected – requiring about 15 hours. Then – because reference material can be helpful – we created a simplified version of the firework control project in a 2 hour video tutorial.

This tutorial is available online. Visit rtv-book.com/chapter16

So, why practice?

1. Software fluency requires maintenance.
2. Repetition and reflection increase skill more quickly than on-gig-experience alone.
3. Established benchmarks provide confidence in timelines and development options.
4. With parametric workflows, assets are adaptable for later use.
5. Exercising creativity alongside practicing skills is an authentic, and efficient, path to becoming a versatile creator.

By Alexandra Hartman, Savages

Case Study Credits
SAVAGES
www.savag.es/

Savages is a real-time content studio based in Los Angeles, CA, focused on screens-graphics, interactive environments, and entertainment experiences.

This studio growth project was supported by:
Ted Pallas, Founder and Principal Designer
Alexandra Hartman, Managing Partner, Designer
Abigail Cinco Aguirre, Designer
Jorge Esquivel Jaime, Designer

CHAPTER 17
Creative First

Real-time content creation is a tool that serves a creative endeavor. The technology that supports real-time content creation presents satisfying challenges while also inspiring creative thought. Experiments with real-time tools can lead to advanced learning, new artistic ideas and new demands on the technical capabilities. When the time comes to build a project for yourself or others, creative should come first. Then choose the right technology to get you there.

Currently, there is a portion of the production community that is churning a lot of hype around real-time content creation. There is a feeling it's cooler than other types of content creation tools. That it's easier to use. That real-time is faster to make content, and cheaper.

These are all myths.

And sometimes they aren't.

These statements can be true if the outlined creative endeavor supports the choice of real-time content creation, which then results in cooler, easier, fast and cheap content production. Those qualifiers don't happen because of a technology choice alone. They happen because good creative direction was accompanied by budget and schedule planning and led by a competent team of producers that communicated clear targets and eliminated barriers to success.

A project was thoughtfully designed, planned, and executed. I will say most shows meet these criteria, and when they do, they still contain challenging, misaligned expectations of what real-time content production tools can achieve. Our job is to eliminate as many misconceptions as possible before accepting a project assignment.

We must demystify the technology and then make difficult and complex tasks feel magical. We earn praise and win projects when we produce work efficiently without involving our clients in the complexity we face. However, we need educated clients who appreciate the complexities we face and then respond with smart budget and schedule choices.

Over time, we will improve our clients' understanding of what helps us succeed in our endeavors. When we start with clear creative goals, we can help our clients achieve sophisticated and expensive endeavors as efficiently as possible.

When Technology Leads Design Discussion

Working professionally with real-time content creation tools, you will often find yourself at the cutting edge of achievable tasks. As we've discussed, these are very new technologies and they can be time-consuming and expensive to use. Workflow communication is still developing. We can't know every detail of the expectations ahead of us, but in this textbook we have outlined tools for good communication.

Sometimes, a client will call wanting to use a particular real-time technology. They might say something like, "I want to do an XR shoot," or, "I want to use Unreal on my next project," or even, "I hear Notch makes cool content."

This client may have seen a video of something that looked totally new and exciting. Invariably they find out what technology made that happen and set out to use the same technology on their own project. This request is technology focused, not design focused. This is a red flag. The tool is not the art.

The tool can inspire artistic thinking. If a client calls with those kinds of statements, follow up with questions like, "what did you see that interested you in that technology?" or "how did you hear about that tool?" See if you can find an existing path to a creative intent.

DOI: 10.4324/9781003206491-20

For example, if a producer believes a tool makes better content and they can't describe what they want that content to look like, that might expose bigger issues with the project. In my experience, the leadership of a tool-driven project rarely ends up happy with the creative result. Leading with technology will not look like the cool thing they saw online. Or worse, the work will take longer, costing more than expected, while the team teases out a sub-par consensus on creative direction.

When the conversation starts with the technical solution, the creative must eventually come from somewhere. This could also become an opportunity to take on a creative leadership role. What inspires you creatively about the technology being discussed? Can that be turned into a creative pitch for the client in response?

What Is the Creative Goal?

New technology is intriguing. There is no doubt a new tool at your disposal will inspire creative thinking. However, new technology will not solve the lack of a creative brief. To properly budget, plan, and produce a project using real-time content creation, there has to be a clear artistic goal. Or else, if the aim is experimentation, I recommend having a flexible deadline (see Figure 17.1).

Most productions live in the center of this Venn diagram where they have a creative target to go with the use of real-time content production. Anything else is going to need a lot of time and/or money to develop the creative. Real-time technology is not currently well suited for creative improvisation on short timelines.

That's not to say experimentation can't be the desired goal. Many teams carve out studio time for artistic experimentation with real-time tools. This can be anything from playing with a Notch block to simulating full XR stages in 3D to experimenting with a new Unreal feature. Experiments in the absence of tight project budgets or deadlines can lead to creative insights that expand your craft.

On a larger scale, when time and budget can afford the process, imagine the satisfaction of a group of incredibly smart and creative people coming together to build something yet undefined with real-time tools. I hope to see more funding to make this possible to move the technology forward. Once the creative juices flow, inevitably the art drives the process forward, pushing to the edge of what the technology can accomplish. That's a good thing!

We find art and technology in this constantly changing relationship that informs both disciplines. Art pushes technological boundaries and technology inspires creativity.

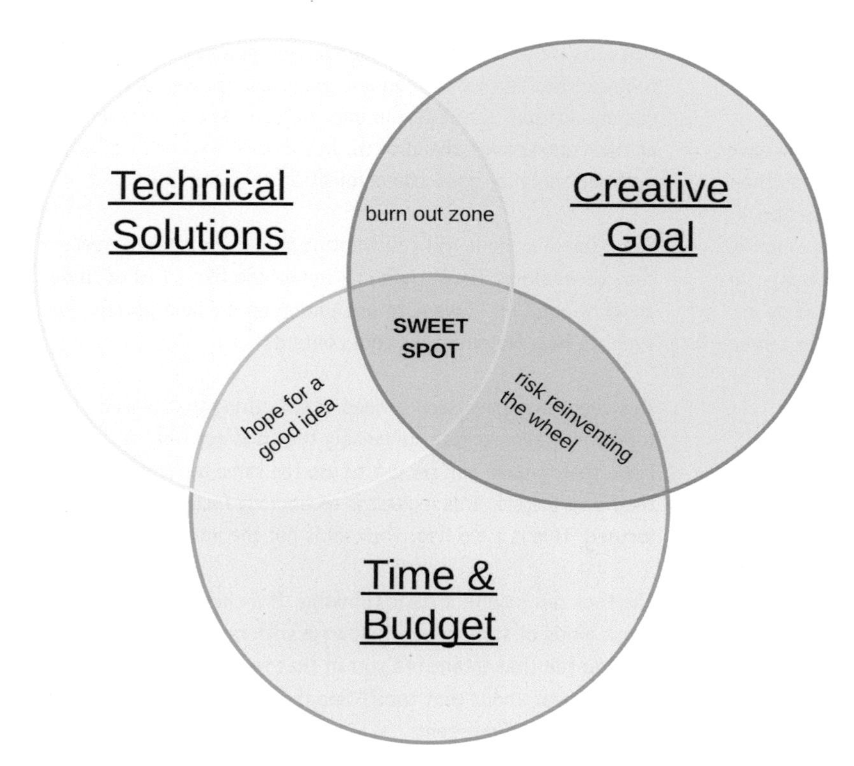

Figure 17.1 Aim for the sweet spot of equal parts creative, tech, and budget. Source: Image by Author

My advice is to beware of projects where this relationship tilts too heavily to tech-driven goals. As a practitioner, I thrive best in the technology and look for creative leadership through my design partners.

Technology Serves Creative Intent

Most of the technologies we use in real-time content creation arose from serving a creative vision. In order to achieve a visual result for art or film or games, many of the technical tools we use began by solving specific creative challenges. The solutions themselves found life beyond the art they initially supported.

These origins stories are from the original founders and early developers of the many tools we commonly use for real-time content today.

> More reading about these platforms and their founders can be found in the companion web page for this chapter at rtv-book.com/chapter17

Matt Swoboda – Founder – Notch

Notch was based on the concepts I developed for the tools that I was using for my own creative projects, demos, etc. with Jani and a few others through the 2000s. We both have a long background in the computer demoscene in the 1990s (the 1980s in Jani's case), which is the original "creative coding." Real-time graphics in demos were used for non-interactive, linear pieces, way back before this was a thing anywhere else.

I'd gone through the iterations of making everything in code (which is as incredibly painful as it sounds) and then building tools to help the process, and then building better tools, and things were starting to get interesting.

We were pretty successful with what we did in the demoscene and attracted some attention and interest for commercial projects outside of that sphere. While there was clearly a gap in quality between what we could achieve with real-time vs offline rendering, the process was much more enjoyable and efficient. It didn't always mean the output had to be in real time. Sometimes it was used for video. It was more about what it meant for the creative process.

In most general cases creative direction should drive technology decisions. In the world of the early real-time graphics (even until pretty recently), this wasn't able to be the case: the technology limits were hard and real, and they very much shaped the creative medium and aesthetic. An entire visual style was developed around those limitations. Amusingly enough, that style is often rehashed in commercial works without any of those limitations today, just because it looks cool. In the beginning, you did what was possible to make the content look good, within the limitations of the hardware. Or ideally the creative bent toward what the limitations of the hardware could do well.

Around the early 2010s, real-time graphics in general started getting close to the threshold where, for a decent number of creative goals, it didn't matter that it was rendered in real time anymore. Real time wasn't a noticeable factor as the image result was "good enough," particularly when the workflow and creative benefits are brought into the equation. There was a degree of luck, I think, that the point where real-time graphics became viable and interesting for wider use happened to coincide with us having the tools and expertise to make use of it.

Ashraf Nehru – Founder – Disguise

D3 came about through simple necessity: the combination of an insurmountable challenge, an impossible deadline, and no other choice.

We (United Visual Artists) had been commissioned to create video content for U2's upcoming Vertigo tour in 2005 and I, as resident software engineer, was responsible for creating and maintaining the custom tools we used. This was an unusual stage design, consisting of seven low-resolution LED video screens in the form of hanging curtains, with pixels the size of tennis balls. The stage was an elliptical design, decorated with linear strips of LED orbiting the center, where the band performed.

The challenge wasn't in creating the content – even though our canvas was just a 150 pixels across, we considered ourselves past masters of the low-resolution aesthetic – but in presenting it to our client, the legendary show designer Willie Williams.

The first meeting, where we presented him the video content for each of the 30-or-so songs as a group of flat rectangles, did not go well. Far from responding positively as we'd anticipated, our client remained neutral and remote throughout. After the meeting, we got the call everyone dreads: "The boss was not convinced. He needs to see the content in three dimensions to make a creative judgment."

This was entirely reasonable, so we set about modeling the stage in 3D Studio, and got to rendering. But we were a small company, with limited rendering resources (also, it was 2005) and by the following week's presentation, we had managed to render out a grand total of five songs. The call afterwards was stark: "The boss was not impressed. You have one more chance – he needs to see all 30 songs in 3D, or . . ."

We had two options (accepting defeat not being one of them): borrow a bunch of powerful computers, or write a real-time 3D stage visualizer. A few phone calls put paid to the first option, so I sat down to write some code. Luckily, I wasn't starting from scratch, but building on a foundation of software tools I'd put together over the last few years – but it was still a close-run thing, with the last bug fixed as our esteemed client rode up in our ancient elevator.

This meeting was different from the others. We were able to show how each song looked, from multiple different perspectives – chosen on the fly, which wouldn't have been possible with rendered video. We could even experiment with different screen arrangements, maximizing the number of seats that could get a good view of the show. The meeting was scheduled to last an hour, but lasted almost three.

After the meeting, we all looked at each other in stunned delight: something important had happened here, something that would forever alter the course of our little company, and of our lives.

Greg Hermanovic – Founder and President – Derivative

It's easy to say we wanted to make a 3D visual/audio software tool for artists that's flexible and general enough that its users could design, experiment, build and perform what they imagined fluently, on-the-fly, in real time. Easy to say, but that's our ongoing mindset.

The key to it all is providing the artist with the space to play freely. Play in manipulating input data, play in image processing, play in 3D, play in building user interfaces and the logic of user interactions, and play in crafting the outputs for your audience.

For me, TouchDesigner grew from a lot of influences prior to it appearing – growing up during the 1960s/1970s electrification of music, watching a lot of 16mm experimental film, toying with modular sound synthesizers, working as a programmer in real-time aerospace simulation, doing visuals for early electronic music culture, and on a larger scale, co-founding Side Effects to make the visual effects products PRISMS and Houdini.

In 2000, Rob Bairos, and Jarrett Smith and I, founded Derivative, soon followed by Ben Voigt, Malcolm Bechard, and Markus Heckmann, at that time "to make innovative tools for designing and performing live visuals for the electronic music culture and for media artists." So our digital roots are in electronic music. The early versions like TouchDesigner 007 and 017 encompassed the idea of users making visual synthesizers (Touch Synths), which they uploaded to our Artworks web page and operated "labels" of synths and remixes.

This was a little ahead of its time so we then went underground in 2005–2008 (well, doing production for Disney Imagineering theme parks, and architects Herzog & de Meuron, and working with Richie Hawtin a.k.a Plastikman) and then we came out with the re-engineered TouchDesigner 077, embedding proper real-time procedural GPU-based 3D and compositing, and pushing deeper in the procedural thing.

Over the years and thanks to a good team of developers, TouchDesigner has been crafted into a professional tool to be deployed in live shows, art installations, video systems, generative art, architecture, XR/VR/AR, visualization, and in schools for teaching.

As developers we want the tools we make to be as flexible as possible in all settings, whether a real-time artist is working solo and controlling all the creative, or working in a small producer/designer/technical team, or at a larger scale in a hand-off process where much of the design happens non-real-time in photoshop, AE, pre-rendered 3D, photography, video, and editing that is transformed and reinterpreted into the real-time work. Hence our obsession with hooking in and inter-operating with other software tools, devices, and protocols – pulling content in easily and pushing new content out.

In real-time projects, once the technical infrastructure is in place, the artist can self-immerse in Play mode – to quickly experiment, improvise, and prototype, where nothing is too precious to abandon or mutate. That's where creativity thrives. We can never undervalue experiment/play/explore time – and it has to happen off-project and on-project. To boost this, TouchDesigner is fully a procedural node-based CG authoring tool which has proven to make artists a lot more productive. Fortunately for the industry, procedural workflows are flourishing in almost all real-time tools now.

Creative software tools these days are getting so feature-rich that a user will never be able to tap into all of it. Which may be good for the artist as the tools are less limiting than ever. But for the tool-developer complexity comes into play, and the challenge is – what's the minimal number of things a user has to learn to give them the maximum amount of creative possibilities? Our mantra includes keeping the software as simple as possible while pushing feature-rich. We continue to learn from what artists across the spectrum are making now, and react to where technology and trends are leading us. But again, easier said than done.

Francis Maes – Founder – SMODE Tech

I started programming SMODE ("Smousse MIDI-Oriented Demo Engine") as a personal project in 2001 just after I realized that I could connect my MIDI controllers to the real-time 3D "demo-making" engine I was programming and that this connection opened a wide new avenue of a "musical approaches to visual content."

This led to a lot of experiments and we decided in 2007 to create a content creation studio D/Labs. We focused on breakthrough real-time content creation, adaptation, compositing, and playback workflow based on SMODE, which we continued to develop as an internal tool. We saw most of the industry using After Effects and suffering from the consequences of the render times, transcoding times, and file transfer time. In our process, all simple operations such as moving a layer, changing a color or making something faster or slower were made in real time, either with a stage simulator or live on the actual stage itself.

With this workflow, D/Labs became the top content creator for French TV Shows and one of the top content creators of concerts in France. With SMODE maturing and becoming effective in a wide number of scenarios, we decided to transform SMODE into a standalone business, SMODE TECH, founded in 2019 by myself and Alexandre Buge, working on SMODE since 2010. "Creative First" has indeed always been at the heart of SMODE and our goal is to create tools to support creative visions.

Emric and Martin – Co-Founders – VYV

The development work VYV has done comes from working with our creative partners, responding to their needs, and our observations on shows through the lens of a development team that understands a cohesive 3D workflow. We always take a few steps back to look for the larger question and consider how to make something that fits in our 3D approach.

Right at the end of their graduate studies, Emric and Martin, with a freshly acquired understanding of 3D Computer Vision, real-time Computer Graphics, and image processing, saw that there was a gap they could fill between what was needed in the real world, state of the art academic research and the rise of computing power provided by programmable GPUs (2004). They knew that the way forward was to create a workflow with 3D comprehension at the core, as anything else would be corrections and compromises (manual warping/offsets and keystone were the standard tools at that time). This approach required the development of their algorithms or building on top of other researchers' work to create tools that could coalesce issues like projection alignment on complex surfaces, lens distortion, and lens shift. The next jump was to create a 3D tracking system that would allow them to compensate for common problems like screen sway, moving set pieces where automation feedback was not available, and surface deformations. This problem was a huge undertaking that required a new open standard for sharing 3D tracking data, and soon, PSN (PosiStageNet) was born.

A few examples include:

- The need to align a multi-projector system covering a volume with multiple automation controlled surfaces, came to developing precise 3D volumetric calibration which reduces the number of calibration points by an order of magnitude (for example, 8x projectors covering a stage and projecting across 10x layers of automation controlled screens requires 8x 10 x 4 = 320 points to line up with automation support, versus 8x8 = 64 points to line up on calibration lines without needing the automation crew to stay all night).
- The need for short load-in times of a touring show with projection mapping came to developing quick and reliable 3D manual alignment tools with simple and tablet-PC-based remote controls (2006), which evolved into a high-speed camera-based auto calibration system around 2013.
- The logistics of deciding infrastructure and projectors placement four years prior to a creation phase on large-scale permanent shows, introduced the use of moving head projectors (flexible but with limited precision on repeatable positioning), which required the development of real-time projector alignment, using attached cameras and IR beacons on stage/screens.
- The need for artistic freedom and re-blocking performer movements with video effects made use of scalable generative effects tied to 3D performer tracking that can span across multiple servers.
- The arms race between lighting designers and projection designers for light output on a single show brought the development of a shared 3D tracking system that can dynamically switch between an automated lighting fixture tracking a performer and a projector replicating the attributes

of that light fixture when they come close to a projection surface so the performer is still lit, but the projection surface is not washed out by the automated light.

- The need for passive performer tracking when trackers could not be used or a skeletal mesh would not capture the expression of a body pose. This need caused us to dive deep into LiDAR technology, finding a way to have multiple LiDAR units work together in a single 3D scene to overcome occlusions.

These founder comments represent a few of the real-time tools we have discussed in this textbook. To learn more about the early history of other platforms visit rtv-book.com/chapter17

Case Study – Luke Halls Studio – Forest of Us

Luke Halls Studio produced the introductory film and interactive visuals for Es Devlin's installation at SUPERBLUE Miami – inaugural exhibition: Every Wall is a Door, 2021.

"Forest of Us takes as its starting point the striking visual symmetries between the structures within us that allow us to breathe and the structures around us that make breathing possible: the bronchial trees that exchange oxygen for carbon dioxide within our lungs and the trees which exchange carbon dioxide for oxygen within our environment."

– Es Devlin

The installation begins with a projected film. Es narrates the concept of her piece which Luke Halls Studio pairs with 3D renders of the forthcoming maze, merging and morphing with forests and bronchial trees. Then when the film finishes, the projection surface opens creating a doorway for guests to enter into the mirror maze (see Figure 17.2), where they can find their reflections distorted and infinitely repeated. When reaching the end of the maze there is a pool of water against the end mirror, where they can see themselves above the water, and as they approach they realize their reflection is filled with branches and as they move out of the mirror's reflection the branches disappear, the trees are housed in their reflection, in their body (see Figure 17.3).

Figure 17.2 Introductory/Entrance Film with the "door" open – projected content surrounds the entrance to the mirror maze. Photo Credit: Andrea Mora.

Figure 17.3 End of the maze interactive visuals. Photo Credit: Alfonso Duran.

At the end of the maze, the mirror content (LED housed in a two-way mirror) isn't real-time generated. Instead a mask triggered and tracked in real time, shaping the content into interactive human form. When a viewer approaches their reflection in the mirror, they find branches that move with their "reflection." The real-time mask is created to fit within a person's reflection with a bespoke build in TouchDesigner (see Figure 17.4). Azure cameras were used to detect when a person stood in front of the mirror and triggered the generation of the live mask.

The team first had to prepare for the vastly varying heights of guests to the installation. The bronchial tree content was initially designed to "attach" to the viewer. Skeleton tracking ensured the content moved with the person when in front of the mirror.

The mirrored space presented another potential challenge for the team (see Figure 17.5). There was concern that IR signals would be reflected around the space causing misfires of the image display. The Azure cameras were dependent on IR for accurate skeleton tracking. We set the planes in front of every mirror and avoided any issues.

We started out with a complicated system based on tracking the human skeleton, using that position data to lock the content to the chest of the person standing in front of the

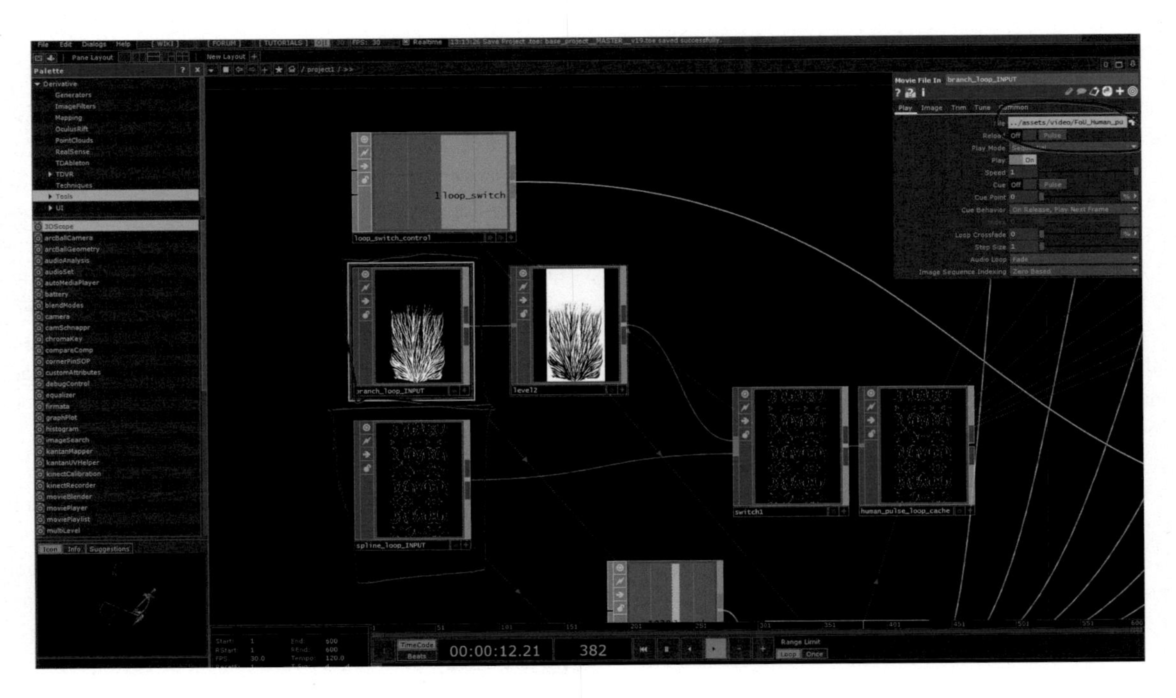

Figure 17.4 TouchDesigner Detail. Source: Image Credit: David Shepherd for Luke Halls Studio.

Figure 17.5 Installation showing LED screens behind ywo way mirror. Source: Image Credit: Luke Halls Studio.

mirror. The branch image followed and responded to the rotation of the viewer. This worked in our testing period, but the result wasn't as elegant as we hoped. We found the content movement wasn't smooth, which was compounded by erratic motion from the viewer adjusting to the content behavior. People engaged with the trick of content tracking their form and less on the meaning of the content; the trees being incorporated into the body.

We also felt that using this effect, the content would be hard to read. The bronchial branches appeared small at a distance over the water trough before the mirror. So we adapted the creative, finding a simpler, almost one size fits all approach. That was achieved by taking an average size person's dimensions at the distance in relation to their size in the mirror and creating content that could cover an entire person, from a central point outward, so that it always filled a person's reflection. Then we designed the content to allow enough negative space for viewers to see their reflection amongst the content.

What we ended up with was essentially a defined front and back plane of space at six points in front of the mirrors. We used the IR and depth cam capabilities of the Microsoft Kinect Azure to analyze a blob and determine when a person had entered the interactivity threshold and activated the pre-rendered content (see Figure 17.6). Their blob (human form) became a luma matte of a looping piece of content. When they stepped out, blobs were analyzed to see whether someone had left and the creative reset for the next participant.

To test our plan, we made a small mock setup with a LCD TV, cameras, and semi two-way flexible mirror. Using this kit the team could work out how we needed to scale and offset the blob feeds in the mirrors at the right distances so that it looked like your reflection, not just a full screen output. Originally we planned to use pressure pads in the floor in front of the mirror to trigger the content but this was removed due to other production reasons. That's when we realized blob analysis would work to trigger the content as well as track the viewer shape. This approach was refined with the small-scale project development setup.

This testing period was also used to determine the camera we wanted to use. Initially we chose the Slim 13e cams (OptiTrack), but they needed to have IR spots at each position as they don't have inbuilt IR, like the Azure cams do. This wasn't possible due to space limitations in the installation so we decided on placing an Azure camera at each guest viewing position.

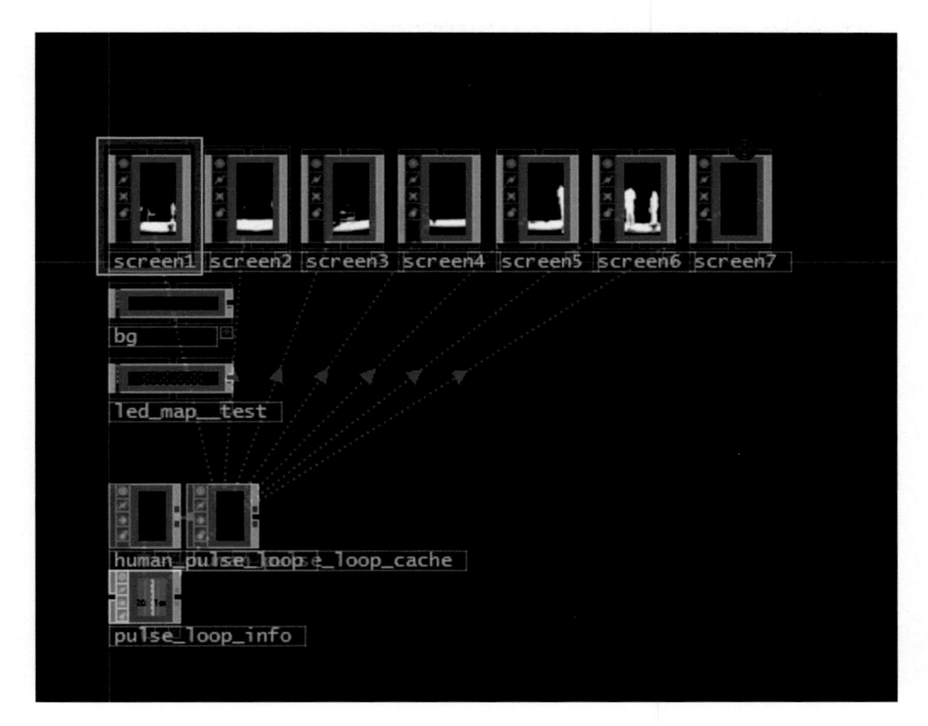

Figure 17.6 TouchDesigner Detail with Blob Tracking. Source: Image Credit: David Shepherd for Luke Halls Studio.

This project was built during the tail end of the Covid-19 pandemic. We had to do all of the fine tuning via zoom calls on laptops, remoting into the servers and looking at the virtual outputs. Not ideal, but we were able to make important refinements working in this manner. Each of the zones were at slightly different distances from the mirror, so the content scale needed to be altered per position for the reflections to match at those different distances. Our final touches were tidying up the smoothing and edge feathering of the mattes to get some of the noise out of the incoming data.

By Charli Davis, Luke Halls Studio

Luke Halls Studio
https://lukehalls.com/

Luke Halls Studio is an award winning, multidisciplinary studio working within projection and video design for the performing arts.

Project Credits
Artist: Es Devlin
Installation Design and Concept: Es Devlin Studio
Video Designers: Luke Halls and Charli Davis
3D Designers: Chris White and Flora Macleod
Interactive Technical Director: Dave Shepherd
Technical Manager: Zakk Hein
Sound Design and Music: Polyphonia Studio

Project name: Forest of Us
Project location: SUPERBLUE Miami
Year: 2021

Appendices

APPENDIX A

Real-Time Content Production Tools and Capabilities

Available: **Y, %, N**
Yes, Depends, No

Complexity Rating: **1-10**
1-low, 9-high

Real-Time Content Technology	Real-Time Content Result					
	Creative Video	Interactive Content	AR	Background Replacement	MR	XR
Custom Code	**Y / 5**	**Y / 6**	**Y / 8**	**Y / 10**	**Y / 10**	**Y / 10**
Media Server *without third party software	**% / 1**	**% / 2**	**N / -**	**N / -**	**N / -**	**N / -**
Real Time Software *utilizing custom hardware	**Y / 3**	**Y / 4**	**Y / 6**	**Y / 7**	**Y / 9**	**Y / 8**
Real Time Platform *media server + RT Software	**Y / 3**	**Y / 4**	**Y / 5**	**Y / 6**	**Y / 7**	**Y / 6**

Figure A.1 This chart represents a shorthand overview of the different methods to produce real-time content, tools used, and relative complexity. Source: Image by Author.

APPENDIX B

VFX Map for Real-Time Content

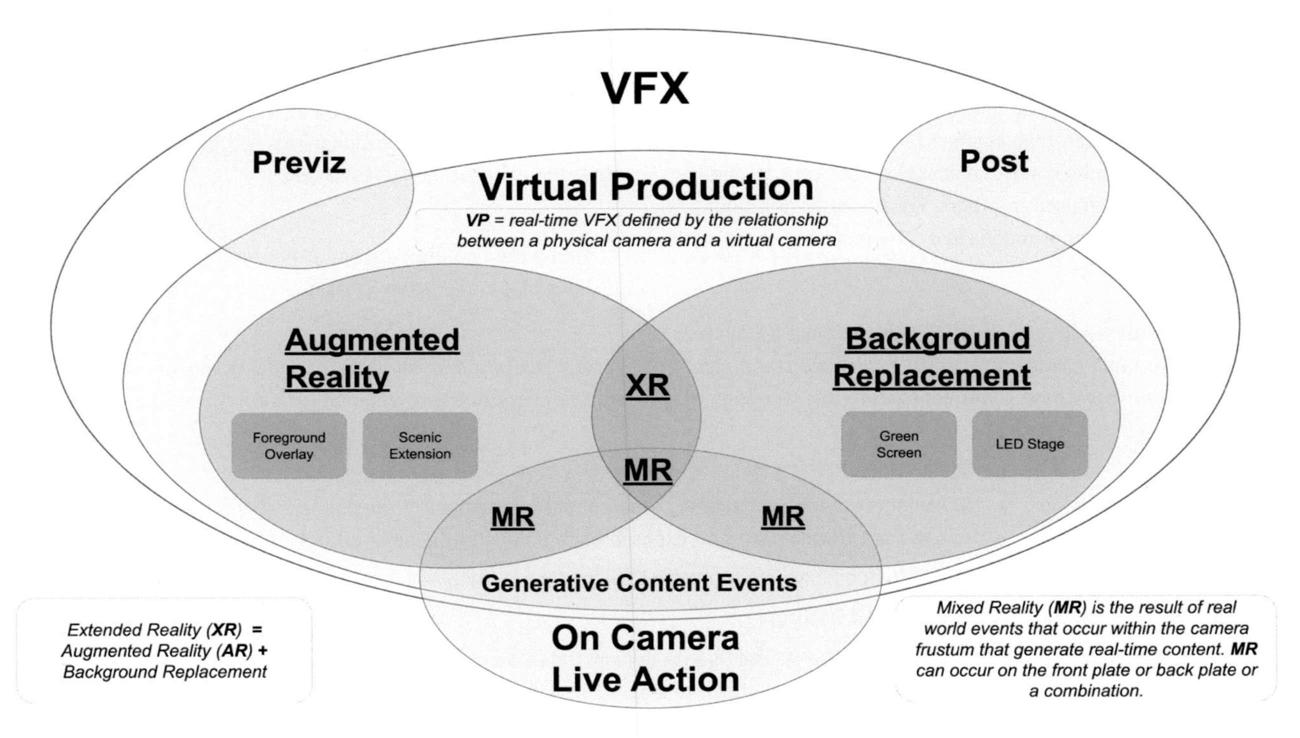

Figure B.1 This diagram outlines the relationship of Virtual Production to visual effects production. Source: Image by Author.

APPENDIX C
Glossary

The following are terms found in this textbook and are related to real-time content production. These terms are defined from their use in the entertainment industry and expand on other industry use when that applies. Terminology is still growing in this discipline and the adaptions of terms in one area of practice are unused in others. Equally true, terms get misapplied, generalized or redefined across practices of real-time content usage.

Communication depends on clear terminology. When the terms are adapting to rapid change, it is easy to misunderstand one another based on how those terms are used in one working environment to the next. Nor can one predict how these terms will evolve in use over time. Please compare this printed glossary with the online version at the companion website for the most up to date definitions. This will be posted along with other industry taxonomy resources.

> Find an online glossary and other terminology reference links at rtv-book.com/glossary

For a Virtual Production glossary under active development be sure to check out www.vpglossary.com/

AI (See Artificial Intelligence)

Ambient Occlusion In 3D computer graphics, modeling, and animation, ambient occlusion is a shading and rendering technique used to calculate how exposed each point in a scene is to ambient lighting.

(from the Wikipedia article https://en.wikipedia.org/wiki/Ambient_occlusion)

AR (See Augmented Reality)

Artificial Intelligence Artificial intelligence is intelligence demonstrated by machines, as opposed to the natural intelligence displayed by animals including humans. AI research has been defined as the field of study of intelligent agents, which refers to any system that perceives its environment and takes actions that maximize its chance of achieving its goals.

(from the Wikipedia article https://en.wikipedia.org/wiki/Artificial_intelligence)

Augmented Reality The general application of Augmented Reality is an interactive experience of a real-world environment where the objects that reside in the real world are enhanced by computer-generated perceptual information. AR can be defined as a system that incorporates three basic features: a combination of real and virtual worlds, real-time interaction, and accurate 3D registration of virtual and real objects.

(from the Wikipedia article https://en.wikipedia.org/wiki/Augmented_reality)

When used in broadcast, Augmented Reality is the compositing of camera sensitive, computer generated images in the foreground of a live camera shot.

As outlined in Chapter 5, four things make Broadcast AR possible:

- Duplication of the camera telemetry from the real world into a virtual computer generated world in real time
- The ability to accurately represent the physical world of the shooting space in a 3D model
- Generation of 3D foreground assets in real time
- Composite of the result with minimal (sub-second) delays

Background Replacement Background Replacement is the real-time generation of camera telemetry responsive, computer generated images in the background of a live camera shot that is either key composited with the camera capture or live captured in the camera background

As outlined in Chapter 5, Background Replacement can be achieved with green screen or LED screen and requires:

- Duplication of the camera telemetry from the real world into a virtual computer generated world in real time
- The ability to accurately represent the physical world of the shooting space in a 3D model
- Generation of 3D background assets in real time
- Option 1: composite of the result with minimal (sub-second) delays

OR

- Option 2A: process imagery for background shot to video output signals
- Option 2B: deliver video signals to video output screen

Back Plate Real-time generated imagery that is destined for the Background Replacement workflow. This background content is upstage of any live action happening within the boundary of the camera frustum and appears live on an LED stage or is composited to the camera shot via a chroma key.

Broadcast AR (See Augmented Reality)

Chroma Key Chroma key is a VFX technique for compositing two images or video streams together based on color. A color range in the foreground footage is made transparent, allowing a separately produced background to be inserted into the scene.

(from the Wikipedia article https://en.wikipedia.org/wiki/Chroma_key)

Compression (See Video Compression)

Creative Coding Creative coding is a type of computer programming in which the goal is to create something expressive instead of something functional. It is used to create live visuals and for VJing, as well as creating visual art and design, entertainment (e.g. video games), art installations, projections and projection mapping, sound art, advertising, product prototypes, and much more.

(from the Wikipedia article https://en.wikipedia.org/wiki/Creative_coding)

Creative Technologist A creative technologist is a person who works in creative technology, a broadly interdisciplinary and transdisciplinary field combining computing, design, art, and the humanities. The field of creative technology encompasses art, digital product design, digital media or advertising and media made with a software-based, electronic and/or data-driven engine. Examples of creative technology include multi-sensory experiences made using computer graphics, video production, digital cinematography, virtual reality, Augmented Reality, video editing, software engineering, 3D printing, the Internet of Things, CAD/CAM, and wearable technology.

(from the Wikipedia article https://en.wikipedia.org/wiki/Creative_technology)

Creative Video Creative video encompasses image production for art and entertainment applications. Examples include media produced for scenic video screens used in rock concerts, theater, television broadcast, and other entertainment design, art installations, artistic displays, and immersive video environments. Creative video can be motion graphic based, augmented shot footage or a combination.

Digital Stage (See LED Stage)

Digital Twin As defined by the Digital Twin Consortium, a digital twin is a virtual representation of real-world entities and processes, synchronized at a specified frequency and fidelity. Digital twins use real-time and historical data to represent the past and present and simulate predicted futures.

(from the Digital Twin Consortium website at www.digitaltwinconsortium.org/2020/12/digital-twin-consortium-defines-digital-twin/)

The game engines used in VP have some digital twin capabilities with customization. Their use in the entertainment industry typically does not require simulation and prediction of the real world or contain the amount of data that would create those simulations. However, this term is being adopted into the VP lexicon to mean a 3D model of an object that has a physical counterpart on or as part of the set.

Downstage In theater, the term downstage applies to any space from the referenced person or object toward the audience. On camera, this term is applied to the space in between the camera and the referenced object or person.

Extended Reality (XR) Extended Reality is the combination of AR and Background Replacement.

Front Plate Real-time generated imagery that is destined for the AR workflow. This foreground content is downstage of any live action happening within the boundary of the camera frustum and is composited over the camera shot. On occasion, background imagery is added to a front plate to extend physical or digital scenery.

Frustum (See Viewing Frustum)

Game Engine A game engine is a software framework primarily designed for the development of video games, and generally includes relevant libraries and support programs. Game engines contain tools available for game designers to code and plan out a video game quickly without building the game world from the ground up. Whether they are 2D or 3D based, these tools aid in asset creation and placement.

(from the Wikipedia articles https://en.wikipedia.org/wiki/Game_engine and https://en.wikipedia.org/wiki/List_of_game_engines)

Genlock Genlock (generator locking) is a common technique where the video output of one source (or a specific reference signal from a signal generator) is used to synchronize other picture sources together. The aim in video applications is to ensure the coincidence of signals in time at a combining or switching point. When video instruments are synchronized in this way, they are said to be generator-locked, or genlocked. This keeps all video capture, playback, and display sources on the same frame.

(from the Wikipedia article https://en.wikipedia.org/wiki/Genlock)

Green Screen (see also VP Green Screen)

Green screen is a type of camera background that employs green for use as a chroma key. Footage is captured on camera, including the green background, which is later removed in post-production or in real time, and replaced with image content.

ICVFX (see In-Camera Visual Effects)

IMAG IMAG, or Image Magnification, is the use of video screens and cameras in a live audience setting to show performers enlarged for higher visibility.

In-Camera Visual Effects In-Camera Visual Effects (ICVFX) refers to the backgrounds and reflections that are captured on camera, rather than added in post-production. The image and reflection sources are typically LED screens that are placed on and off camera.

Key (See Chroma Key)

LED Stage Any configuration of LED screen tiles to surround the live action of performers and physical presence of props and scenery. When used in Virtual Production, the LED stage provides background and reflections from imagery generated as a back plate from real-time content source.

LiDAR LiDAR is an acronym of "light detection and ranging" or "laser imaging, detection, and ranging." It is sometimes called 3D laser scanning, a special combination of 3D scanning and laser scanning. LiDAR is a method for determining ranges (variable distance) by targeting an object or a surface with a laser and measuring the time for the reflected light to return to the receiver. It can also be used to make digital 3D representations of objects or surfaces.

(from the Wikipedia article https://en.wikipedia.org/wiki/Lidar)

LiDAR creates a point cloud representing the measurement data of the reflected laser light. That point cloud must be converted to a polygonal model for use in projection mapping, AR, and LED stage workflows.

Lossless Compression (See Video Compression)

Lossy Compression (See Video Compression)

Luma Matte A luma matte is a black and white reference to define transparency when compositing two video signals together. Black is defined as opaque and white is defined as Alpha or transparent.

Machine Learning Machine learning is a field of inquiry devoted to understanding and building methods that "learn," that is, methods that leverage data to improve performance on some set of tasks. It is seen as a part of artificial intelligence. Machine learning algorithms build a model based on sample data, known as training data, in order to make predictions or decisions without being explicitly programmed to do so.

(from the Wikipedia article https://en.wikipedia.org/wiki/Machine_learning)

Machine learning algorithms are used in real-time content creation to support facial tracking and procedural content generation.

MediaOps (Media Operations) MediaOps refers to the workflow that manages creative video content production. It is the responsibility of the MediaOps team to outline content delivery requirements, assist with previsualization and run playback on production site.

Metaverse The term "metaverse" was coined by Neil Stephenson in the 1992 novel *Snow Crash*. It is used in the book to describe a virtual world accessed by VR headsets. More recently, it is used in the 2020s to describe the growing 3D social and gaming networks that are typically viewed on a flat screen. As live entertainment looks to accommodate a hybrid of in-person and virtual audience attendance, metaverse solutions will continue to borrow from VP technologies that are combined with cloud infrastructure to realize the next generation of online experiences.

Mixed Reality (MR) Mixed Reality is the result of real world events that occur within the camera frustum that generate real-time content. MR can occur on the front plate or back plate or a combination.

As VP is real-time content generated by the real world action of moving the camera, it is important to make the distinction that MR is part of the live action captured on camera.

ML (See Machine Learning)

MR (See Mixed Reality)

Pixel In digital imaging, a pixel or picture element is the smallest addressable element in a raster image, or the smallest addressable element in an all points addressable display device; so it is the smallest controllable element of a picture represented on the screen. Each pixel is a sample of an original image; more samples typically provide more accurate representations of the original. The intensity of each pixel is variable. In color imaging systems, a color is typically represented by three or four component intensities such as red, green, and blue, or cyan, magenta, yellow, and black.

(from the Wikipedia article https://en.wikipedia.org/wiki/Pixel)

POV (See Point of View)

Point of View (POV) Point of view is used to refer to the camera or audience member viewing position. The viewing position accounts for location and angle relative to the viewed scene. When a camera POV is described, the data points will include zoom status to account for field of view.

Previsualization (Previs or Previz) Previsualization entails creating a 3D model of a scene to represent the physical result of a proposed construction. When used in entertainment production, this could be a set design, a lighting design, a scenic screen video design, a camera move, and so on. Previz can be viewed on a self-contained hardware + software platform, or be cloud based for remote viewing options.

Previs or Previz (See Previsualization)

Projection Mapping Projection mapping is a process in which a video is projected onto a dimensional surface. The map includes correctly associating a pixel with a location in space, like a point on a building. Video content is also "mapped" onto LED screen surfaces by a similar logic.

Ray Tracing In 3D computer graphics, ray tracing is a technique for modeling light transport for use in a wide variety of rendering algorithms for generating digital images.

(from the Wikipedia article https://en.wikipedia.org/wiki/Ray_tracing_(graphics))

Real-Time Video content that is rendered in real time is typically referred to as real-time content, and shortened to the term, "real-time."

Note that there are many instances in the textbook of the combination of the two words, "real" and "time," hyphenated and unhyphenated. To know what usage is appropriate, check if you are using "time" as a noun or as part of an adjective to describe something. An event happening "in real time" is a different usage than a "real-time event." When "real-time" is used in isolation, the item being described is often dropped, and should include the hyphen.

Scenic Extension Scenic Extension is a process by which the portion of the back plate visible on an LED stage is extended into an AR front plate. When done correctly, the background looks seamless from the physical LED wall to AR overlay.

Screens Producer A Screens Producer is an individual who monitors the life of the video pixels from their creation to their delivery on a video screen. This role covers content design to system engineering, providing a team to oversee infrastructure, playback, content creation, and delivery workflow.

SMPTE timecode SMPTE timecode is a set of cooperating standards to label individual frames of video or film with a timecode. The system is defined by the Society of Motion Picture and Television Engineers in the SMPTE 12M specification. Timecodes are added to film, video, or audio material, and have also been adapted to synchronize music and theatrical production. They provide a time reference for editing, synchronization, and identification. Timecode is a form of media metadata. The invention of timecode made modern videotape editing possible and led eventually to the creation of non-linear editing systems.

(from the Wikipedia article https://en.wikipedia.org/wiki/SMPTE_timecode)

Subsurface Scattering Subsurface scattering, also known as subsurface light transport, is a mechanism of light transport in which light that penetrates the surface of a translucent object is scattered by interacting with the material and exits the surface at a different point. The light will generally penetrate the surface and be reflected a number of times at irregular angles inside the material before passing back out of the material at a different angle than it would have had if it had been reflected directly off the surface. Subsurface scattering is important for realistic 3D computer graphics, being necessary for the rendering of materials such as marble, skin, leaves, wax, and milk. If subsurface scattering is not implemented, the material may look unnatural, like plastic or metal.

(from the Wikipedia article https://en.wikipedia.org/wiki/Subsurface_scattering)

Telemetry Telemetry is the collection of predefined data points that are real-time processed. Camera telemetry used in VP relates to data regarding the camera's POV and includes 3D position, lens direction, zoom status, shutter angle, and shutter speed.

Upstage In theater, the term upstage applies to any space from the referenced person or object away from the audience. On camera, this term is applied to the space in between the backdrop and the referenced object or person.

VAD (See Virtual Art Department)

VADOps (See also MediaOps and XROps)

The VADOps (or Virtual Art Department Operations) team is responsible for the delivery of imagery to the LED stage screen environment. This includes documenting delivery workflow, managing digital assets, integrating with real-time data sources, and the engineering and delivery of real-time content signals.

Video Compression In information theory, data compression, source coding, or bit-rate reduction is the process of encoding information using fewer bits than the original representation. Any particular compression is either lossy or lossless. Lossless compression reduces bits by identifying and eliminating statistical redundancy. No information is lost in lossless compression. Lossy compression reduces bits by removing unnecessary or less important information. Typically, a device that performs data compression is referred to as an encoder, and one that performs the reversal of the process (decompression) as a decoder.

(from the Wikipedia article https://en.wikipedia.org/wiki/Data_compression)

For a list of video compression types for rendered files, check the Wikipedia article on Video Codecs (https://en.wikipedia.org/wiki/Video_codec). Real-time content is transmit across a video signal distribution system that may be compressed depending on transmission type.

Viewing Frustum In 3D computer graphics, the viewing frustum is the region of space in the modeled world that may appear on the screen; it is the field of view of a perspective virtual camera system. The view frustum is typically obtained by taking a frustum – that is a truncation with parallel planes – of the pyramid of vision, which is the adaptation of a cone of vision that a camera or eye would have to the rectangular viewports typically used in computer graphics.

The exact shape of this region varies depending on what kind of camera lens is being simulated, but typically it is a frustum of a rectangular pyramid.

(from the Wikipedia article https://en.wikipedia.org/wiki/Viewing_frustum)

Virtual Art Department The Virtual Art Department is responsible for the creation and maintenance of digital assets for use in the real-time content creation workflow.

Virtual Production (VP) Virtual Production is a real-time VFX defined by the relationship between a physical camera and a virtual camera. The virtual camera exactly mimics the attributes and behavior of a physical camera in a simulated 3D environment, generating perspective sensitive content separated into foreground and background plates.

The background plate is fed to different Background Replacement formats including: physical screen systems for ICVFX capture or composited into the physical camera feed using green screen/background key. The foreground plate is composited into the physical camera live feed for AR and scenery extension.

Volume (see also LED stage)

A volume is the preferred terminology for an LED stage used in a film shoot. These stages are large curved walls, and sometimes fully enclosed by LED screen and have LED panels in the ceiling. The floor is treated with practical ground covering.

VP (See Virtual Production)

VP Green Screen VP green screen is a special case of the use of green screen combined with VP tools, specifically the real-time generation of camera sensitive background content that is composited to the live signal.

VP Producer (See also Screens Producer and XR Producer)

A VP Producer is an individual who monitors the life of the video pixels in a real-time working environment from their creation to their delivery to a video screen. This role covers content design to system engineering, providing a team to oversee infrastructure, real-time data input, playback, content creation, and delivery workflow.

UV Mapping UV mapping is the 3D modeling process of projecting a 2D image to a 3D model's surface for texture mapping. The letters "U" and "V" denote the axes of the 2D texture because "X," "Y," and "Z" are already used to denote the axes of the 3D object in model space, while "W" (in addition to XYZ) is used in calculating quaternion rotations, a common operation in computer graphics.

(from the Wikipedia article https://en.wikipedia.org/wiki/UV_mapping)

Virtual Reality (VR) Virtual reality (VR) is a simulated experience that can be similar to or completely different from the real world. Applications of virtual reality include entertainment (particularly video games), education (such as medical or military training) and business (such as virtual meetings).

(from the Wikipedia article https://en.wikipedia.org/wiki/Virtual_reality)

VR (See Virtual Reality)

XR (See Extended Reality)

XROps (see also MediaOps and VADOps)

The XROps (or XR Operations) team is responsible for the delivery of imagery to the LED stage screen environment. This includes documenting delivery workflow, managing digital assets, integrating with real-time data sources, and the engineering and delivery of real-time content signals.

XR Producer (See also Screens Producer and VP Producer)

A XR Producer is an individual who monitors the life of the video pixels in a real-time working environment from their creation to their delivery to a video screen. This role covers content design to system engineering, providing a team to oversee infrastructure, real-time data input, playback, content creation, and delivery workflow.

XR Stage (See also LED stage)

XR stages are a type of LED stage that is commonly used in an XR workflow. The floor will often be composed of LED screen and the ceiling is commonly used for live entertainment type lighting rigs.

Index